£12·00

OECD PROCEEDINGS

ENVIRONMENTAL BENEFITS FROM AGRICULTURE: ISSUES AND POLICIES

The Helsinki Seminar

PUBLISHER'S NOTE

The following texts are published in their original form to permit faster distribution at a lower cost.
The views expressed are those of the authors, and do not necessarily reflect those of the Organisation
or of its Member countries

ORGANISATION FOR ECONOMIC CO-OPERATION AND DEVELOPMENT

ORGANISATION FOR ECONOMIC CO-OPERATION AND DEVELOPMENT

Pursuant to Article 1 of the Convention signed in Paris on 14th December 1960, and which came into force on 30th September 1961, the Organisation for Economic Co-operation and Development (OECD) shall promote policies designed:

- to achieve the highest sustainable economic growth and employment and a rising standard of living in Member countries, while maintaining financial stability, and thus to contribute to the development of the world economy;
- to contribute to sound economic expansion in Member as well as non-member countries in the process of economic development; and
- to contribute to the expansion of world trade on a multilateral, non-discriminatory basis in accordance with international obligations.

The original Member countries of the OECD are Austria, Belgium, Canada, Denmark, France, Germany, Greece, Iceland, Ireland, Italy, Luxembourg, the Netherlands, Norway, Portugal, Spain, Sweden, Switzerland, Turkey, the United Kingdom and the United States. The following countries became Members subsequently through accession at the dates indicated hereafter: Japan (28th April 1964), Finland (28th January 1969), Australia (7th June 1971), New Zealand (29th May 1973), Mexico (18th May 1994), the Czech Republic (21st December 1995), Hungary (7th May 1996), Poland (22nd November 1996) and the Republic of Korea (12th December 1996). The Commission of the European Communities takes part in the work of the OECD (Article 13 of the OECD Convention).

Publié en français sous le titre :

AVANTAGES ÉCOLOGIQUES DE L'AGRICULTURE : ENJEUX ET STRATÉGIES
Le séminaire d'Helsinki

FOREWORD

The Seminar on Environmental Benefits from Agriculture, hosted by the Finnish Ministry of Agriculture, was held in Helsinki on 10-13 September 1996, and included a one-day study visit to farms implementing the European Union's agri-environmental measures. It was opened by the Finnish Minister of Agriculture, and brought together around 100 participants from agriculture and environment ministries in 22 OECD countries and 1 OECD Observer country, 4 international non-governmental environmental organisations and representatives of farmers professional organisations. The Finnish Minister for Environment hosted a lunch during the Seminar.

The OECD Secretariat presented an overview paper, as the basis for discussion of four consultant papers on conceptual areas on the nature and measurement of environmental benefits from agriculture and a set of country case studies, prepared by the countries themselves, describing specific policies and practices addressing the issue of environmental benefits from agriculture. The participation of non-governmental environmental organisations, researchers, and professional farmers organisations was particularly useful in widening the area of discussion.

The Seminar was an integral part of the programme of work on agri-environmental issues in the OECD. Its purpose was to define the key policy issues and the experience and role of different policy measures and market solutions in OECD countries, in the context of agricultural policy reform, and to suggest areas that might require further work to be undertaken in the OECD. There was a very useful general discussion based on an oral summary of the content and the discussions of the papers, introduced by the rapporteur. This led to a broad consensus on the conclusions that emerged from the Seminar.

The Joint Working Party of the Committee for Agriculture and the Environment Policy Committee discussed the summary and the conclusions of the Seminar in December 1996. Subsequently, the Committee for Agriculture and the Environment Policy Committee agreed to recommend the derestriction of this document on the responsibility of the Secretary-General. The conclusions of the seminar, the overview paper from the OECD Secretariat, the Summary by the rapporteur, the conceptual papers prepared by consultants, the official statements, and summaries of the country case studies, prepared by the countries themselves, are included in this publication.

Fifteen countries prepared country case studies outlining their policies and experiences relating to environmental benefits of agriculture. The full set of country case studies is available separately on INTERNET, as a general distribution document. The list of the country case studies corresponds to the list of summaries included in the Table of Contents of this publication which is published on the responsibility of the Secretary-General of the OECD.

The OECD expresses its appreciation to the Finnish authorities for the very active role they played in preparing, arranging and hosting the Seminar, as well as to eight other financial contributors: Belgium, the European Commission, Germany, Japan, the Netherlands, Norway, Switzerland and the United States.

TABLE OF CONTENTS

CONCLUSIONS OF THE SEMINAR

Introduction

OECD countries are committed to the reform of agricultural policies and the economically and environmentally sustainable management of natural resources. There has been progress in reforming agricultural policies, with some shift away from market price support, towards the use of direct payments and other budgetary support to agriculture, together with greater opening of domestic markets to world markets. However, the extent and pace of reform varies widely across OECD countries. The reform process has been substantially progressed by the conclusion of the Uruguay Round agreement on agriculture.

It is against this background that policy makers in OECD countries are addressing the issues related to improving environmental quality and performance in agriculture. In this context, work in the OECD is seeking to identify ways in which governments might promote market solutions and, in addition, design and implement policies to achieve environmentally and economically sustainable agriculture, with least resource cost and trade distortion.

Agricultural activities have both beneficial and harmful effects on the environment through changing the quality or quantity of soil, water, air, natural habitats, biodiversity and landscapes. The extent of these environmental effects depends on the effects of farming practices on ecosystems at the local, regional, national and international levels. While some of the environmental effects of agriculture occur off the farm, it can sometimes be difficult to attribute particular impacts to specific farms.

In some OECD countries the development of agriculture is weakening the coupling of land management associated with biodiversity and sustainable development, which were related to a variety of historical agrarian systems.

Farmers have an incentive to adopt farm practices that maintain the quality and quantity of the natural resources, in so far as reducing the value of these resources will impede their operations and viability in the future. But this incentive depends on farmers being aware of the environmental costs and benefits of their activities, and that markets reflect these effects, which are influenced by agricultural and environmental policies.

There is a considerable diversity among the OECD countries in their agricultural systems and in the nature of the environmental attributes associated with their farming practices. Therefore, no single policy solution would be appropriate. The policy challenge is nevertheless to reduce the harmful environmental effects and enhance the beneficial effects, including through specific environmental policy measures in agriculture, which are consistent with agricultural policy reform.

An important aspect of the agricultural policy reform process is to facilitate opportunities for farmers to develop alternative source of income. One way is for farmers to provide rural amenities, which are very diverse in nature, and are also provided by other sectors of the rural economy.

1. Main questions

The seminar brought together government officials, representatives from professional farm organisations, environmental groups and researchers, who focused on a number of key questions:

- What are the characteristics of environmental benefits, or services, from agriculture?

- What is the variety of environmental benefits from agriculture perceived and demanded by society across OECD countries, and how is their provision related to agricultural policies?

- How can non-market environmental benefits be valued?

- What has been the experience of OECD countries in addressing the environmental benefits from agriculture through policy measures and market approaches, and how have the policies been evaluated and monitored? How can the evaluation and monitoring of policies be improved and which indicators are needed for such an evaluation?

- How can consistent agricultural and environmental policies be designed and implemented so that both the environmental benefits and harm due to agricultural activities are dealt with in a balanced way?

- In which areas might work in the OECD contribute to the further understanding of the policy issues involved in addressing the environmental benefits from agriculture, and in providing guidance for policy makers?

Several conclusions were reached in the course of a wide-ranging and fruitful discussion of these questions, and other policy issues.

2. Main conclusions

Conceptual

There is a range of environmental effects which are perceived as environmental benefits or services from agriculture. For some countries these benefits are understood to only include biophysical and ecological environments, such as unique flora and fauna which have evolved with low-intensity farming over time, and which in some cases are threatened by current systems of agricultural production. For other countries they also include significant attention to landscape and cultural features, as well as considering rural development aspects of agriculture, which are intended to contribute to balanced national and regional development.

With respect to each of these agri-environmental aspects, there is a general understanding of the need for increasing the awareness of farmers and the public of the environmental effects of farming, and for a reference level (or benchmark) that distinguishes beneficial effects from harmful effects. While the reference level should be based on the environmental outcome of good farming practice, it varies within and across countries, and over time, and is influenced by different perceptions of agricultural sustainability.

The environmental services from agriculture are jointly produced from farming practices, and when a market exists to remunerate these services the market itself will bring forth environmental benefits, and contribute to farmers' income.

However, where markets are incomplete there is a difficulty in reflecting and transmitting to farmers the appropriate level of environmental benefits demanded by society. In these cases collective action may be justified, which is an issue for public policy.

Policy framework

In some OECD countries there are policy issues concerned with the importance of rural amenities provided within specific areas, yet valued by society as a whole, but not fully reflected in rural incomes.

When public policy is justified, the combination of agricultural and environmental policy measures need to be carefully designed and implemented in order to ensure that they improve rather than worsen the environment compared to a situation in the absence of public policy. Measures should be designed to avoid reinforcing perceptions that existing property rights are absolute and inviolable, which could complicate the achievement of environmental policy objectives in the longer term.

There is a general recognition that agricultural policy reform, through lowering both output and input linked-support to agriculture, will contribute to providing the right incentives to farmers with respect to environmental concerns. However, reform may not be sufficient to achieve the level of desired environmental outcomes, in part because markets are incomplete for some environmental benefits.

There is a wide-range of agricultural and environmental policy measures and approaches which have been implemented across OECD countries with the aim of improving the environmental performance in agriculture. However, sometimes the effect of these measures to improve environmental performance has been compromised where they have been implemented together with policies to support agriculture. This arises because some agricultural policies encourage farming practices which have contributed to environmental damage.

Within OECD countries many of the policy measures and approaches to address the environmental benefits of agriculture have only recently been introduced and, in general, their effects have not yet been assessed. A wide array of approaches are available, ranging from local, regional and voluntary approaches; the dissemination of the results of research; education and training; regulatory measures; and financial incentives and disincentives to farmers. However, these approaches may have different effects on the environment and it will be necessary to evaluate and monitor their overall effects.

Ultimately, it is the actions and performance of farmers themselves that will bring about the environmental effects from agriculture. A key imperative is that policy measures give the right incentives to encourage and not impede improved environmental performance by farmers. In this regard, drawing up codes of good agricultural practices can help farmers to adopt sustainable farming practices.

Payments to farmers

Many of the policy measures currently implemented in OECD countries provide payments to farmers for the provision of environmental benefits. However, while some of these payments are well targeted, others are not, and are implemented without an overall evaluation of the associated costs in relation to the environmental benefits. In some cases the payments are often applied uniformly at national or regional level, yet the benefits are locally concentrated or site specific. Therefore, there can be an important role for regional authorities to meet regional needs and local targets.

Payments should only be used where necessary to achieve a programme's environmental objective, taking account of available alternatives. To the extent possible, payments should be linked to environmental outcomes, or farming practices which determine those outcomes, rather than the volume or type of production, or factors of production not directly related to the environmental outcome. Payments should also be available to anyone who can provide the environmental benefits in question.

In addition, in order to ensure that any payments are cost effective in achieving their objectives to provide environmental benefits, and do not distort agricultural markets, they need to be:

- transparent in their objectives and operation;

- targeted to ensure the provision of the benefits, which would not be otherwise provided above the recognised reference level;

- tailored to particular environmental situations, limited to cover the costs of providing the benefits, and accompanied by adequate advice and information;

- evaluated as to their environmental effects, the results of which would feed back into the possible adapting of programmes to ensure that they meet environmental needs through alternative lower cost solutions;

- monitored to ensure compliance and cost effective implementation.

While there is an impressive body of knowledge, both concerning the conceptual aspects of the issues and the practical possibilities to improve environmental performance in agriculture, this knowledge has often not been sufficiently applied nor provided and disseminated to farmers. Thus there can be a role for governments to implement policies and programmes designed to raise levels of farmers awareness of sustainable agricultural practices.

Payments should also be based on sound science of physical and biological processes to ensure that farming practices attain desired outcomes and avoid unintended damages.

————————————

A greater understanding of the nature, experience and environmental effects of policies is required. In particular, the development of appropriate indicators at national and regional level can contribute to the assessment of performance and the evaluation of policy measures. In the analysis and evaluation of policy measures across countries and the development of appropriate indicators, there is an important role for the OECD.

In designing and implementing policy measures to enhance the sustainability of agriculture and improve environmental performance, there is a general understanding that both the beneficial and harmful environmental effects of agriculture should be considered. In this context, OECD countries are committed to apply the Polluter Pays Principle (PPP) to avoid negative effects across all sectors including agriculture. Although there may be some difficulty in implementing the principle in agriculture, efforts should be made to move in that direction.

3. Implications for future work

The Seminar contributed an impressive amount of information and analysis on underlying conceptual issues and country experiences, for the ongoing work on agricultural policy reform and the environment, and on developing agri-environmental indicators. The Seminar also provided further information on agri-environmental measures, of use in the analysis and monitoring of agricultural policies.

However, the Seminar also made clear the need for further work to help policy makers to design and implement relevant policies in an operational way, and to promote market solutions to achieve environmentally and economically sustainable agriculture at least resource cost to the economy and with least trade distortions. There is still a need to develop some of the conceptual issues, the analysis of the linkages between agriculture and environmental benefits, and the measurement of environmental benefits themselves. Work is also needed to explore the role of policy, the use of alternative policy instruments and approaches, and the role of market solutions to address these issues.

Some of this work might best be pursued in national administrations, while some might best be undertaken in the OECD. Further work in the OECD could include the following main areas:

- Valuation of environmental costs and benefits, including through the development of appropriate indicators.

- Analysing the concept and nature of the "reference level" which can help to distinguish between beneficial and harmful environmental effects of agriculture.

- Monitoring the budgetary costs and outcomes of policy measures that provide payments for the provision of environmental benefits to agriculture in OECD countries.

- Exploring ways of promoting efficient solutions to enhance environmental benefits and reduce environmental harm, bearing in mind the various international agreements in the field of the environment, which can have implications for agriculture.

- Identifying the main operational criteria in the design and implementation of policy measures, especially budgetary payments for the provision of environmental benefits of agriculture, which are compatible with agricultural policy reform (this could extend the study on the characteristics of direct payments given in *Agricultural Policy Reform: new approaches - the role of direct income payments*. OECD, Paris 1994).

OVERVIEW AND MAIN POLICY ISSUES: AN OECD PERSPECTIVE

by
Luis Portugal
Directorate for Food, Agriculture and Fisheries
OECD, Paris

Introduction

The natural environment supplies the basic resources for agriculture activities and is shaped by those activities. There is a close, complex and dynamic relationship between natural resources used in agriculture and the environment, and agriculture is a sector (with forestry) in which production activities can generate both beneficial and harmful effects on the environment. On the one hand, sustainable farming systems can be associated with the maintenance of traditional landscapes, the preservation of natural habitats and biodiversity, water and soil management, and the sustaining of rural communities and cultures. On the other hand, agricultural activities can lead to the pollution of surface and ground water by fertilisers, residues of farm chemicals, and nitrates and phosphates from livestock manure; soil and water problems due to tree felling to increase agricultural land; visual pollution from unsightly farm buildings, and air pollution from air-borne residues, especially from livestock enterprises.

A number of agricultural policy reform efforts are under way with a view to lowering the levels of overall support to agriculture and shifting policy measures away from market price and other commodity linked support and towards budgetary financed measures, which have the potential to be better targeted and less production and trade distorting. Budgetary support to farmers for environmental purposes is increasingly becoming part of the policy mix in many OECD countries. Such support is often intended to compensate farmers for undertaking specified activities deemed either to "enhance" environmental benefits or to "reduce" harmful environmental effects or to "prevent" further harm to environment. Some measures are conditional upon the farmers undertaking or changing certain agricultural practices, or on withdrawing land from agricultural production.

In some countries low budget cost solutions are preferred to encourage voluntary, co-operative efforts among farmers, or to encourage the dissemination of information and training to farmers. Public financing of structural and infrastructural projects has been provided in some countries to avoid the abandonment of agricultural production or to preserve some environmental beneficial farming systems in certain areas.

Yet in most OECD countries, agriculture remains a heavily assisted industry, with support policies having multiple and sometimes contradictory effects on the environment. Many of these measures, which provide relatively high levels of support and are to varying degrees coupled with production of specific agricultural commodities or input use, encourage production and practices which put pressure on environment. Yet other measures are implemented at the same time, intended to offset the harmful environmental effects of such measures. This raises questions as to their efficacy in achieving their environmental goals with least production, trade, and economic distortions, and the extent to which they are consistent with the Polluter Pays Principle.

While the focus of the Seminar is on environmental benefits from agriculture - the nature, extent and evaluation for policy purposes - nevertheless it is important to recognise the broader policy aims to which OECD Ministers are committed, which is the reform of agricultural policies to ensure and enhance the sustainability of agriculture. The overall objective of the OECD work on agriculture and the environment is to analyse policy and market solutions to improve environmental performance of agriculture, and suggest ways in which governments might design and implement policies that contribute to enhancing sustainable agriculture at minimal resource cost to the economy and with least production and trade distortions.

The OECD's work on agriculture and the environment is being developed within the broader context of the commitment to agricultural policy reform made by OECD Ministers in 1987. The main thrust of the drive for this reform is to let market signals influence the orientation of agricultural production, through a progressive reduction of support and protection of the agricultural sector. While pursuing this long-term objective, consideration may also be given to other concerns, including environmental protection. Rather than being provided through price guarantees or other measures directly linked to production or to factors of production, farm income support should be sought, as appropriate, through direct payments. These principles were reaffirmed in 1992 by the meeting at Ministerial level of the Committee for Agriculture of the OECD, which also stressed the growing importance of the mutual relationships between agriculture and the environment.

After the successful conclusion of the Uruguay Round in December 1993, the OECD Committee for Agriculture met at High Level in February 1994 and expressed the need to further explore the role of the various policy instruments and to define their most appropriate combination, or "policy mix", in advancing the reform process. It was recognised that, whilst market-oriented reform is generally expected to bring environmental benefits, specific complementary measures targeted to desired environmental outcomes will also be necessary. Furthermore, the adjustment of the agricultural sector will be facilitated if it is supported by comprehensive policies for the development of various activities in rural areas, notably to help farmers to find supplementary or alternative income.

The Seminar can provide an important contribution to the policy dialogue within this broad policy context in an OECD perspective by:

- examining the nature and evaluation of environmental benefits from agriculture;

- sharing experiences of policies measures and approaches, and farmers own actions addressed to enhancing environmental benefits from agriculture across the range of OECD countries;

- focusing the discussion within the overall context of agricultural policy reform, sustainable agriculture, and the need to take a coherent approach to agricultural and environmental policies.

1.   Agricultural activities, agricultural policies and environmental effects

The major linkages between agricultural activities and the environment derive from the effects of farming practices on:

- agricultural land use and soil quality, including the cultivation of marginal land;

- water management;

- air quality;

- the diversity of animal and plant species and the preservation of wildlife habitats and ecosystems;

- the natural landscape; and

- the rural development.

It is important to note that all of these effects of agricultural activities can be either "positive" (beneficial) or "negative" (harmful) to the environment depending on the direction of change. Moreover, the judgement as to what are beneficial and harmful effects is generally subjective and there is often no, or only incomplete, markets which reflect preferences by the society, and incomplete establishment of property rights. Even when such a judgement can be established, there are often serious problems of valuation, because agricultural policies have changed the base line against which changes may be evaluated: many policies have been in place for a long time, the environmental effects take time to appear, and are not always farm specific.

Agricultural policies, especially output- and input-related support, have in the past intensified farm production and have contributed to the environmental problems in agriculture, to varying extents across commodity sectors, regions and countries. The impacts of agricultural policies on the environment are essentially the result of the effects of changes in policy measures on the scale and location of farming, and farming practices. Agricultural policies influence farming activities by changing the incentives in employing resources in agricultural activities, in particular through:

- changes in output and input prices;

- restrictions on output and in input use;

- (dis)incentives for the development and the adoption of new technologies and practices;

- removal or creation of impediments to resource movement;

- changes in the agricultural and rural infrastructure;

- provision of information and advice, training, research and development.

In addition, environmental policies also influence farming activities that affect the environment by:

- placing regulations on certain activities (especially concerned with chemical use, waste disposal and farm practices);

- providing taxes or charges on certain farming activities;

- influencing the "climate of public opinion".

While farmers adjust their decisions in response to changes in incentives as a consequence of policies, these decisions tend to only include the weighing up of private costs and benefits. Agricultural policy reform will lead to reductions in the level of support and thus changes in relative prices between commodities, regions and countries, farm inputs and outputs. In their decisions on how allocate their resources, farmers will increasingly face the changed incentives offered by markets, as well as a different set of risks and uncertainties. This will result in changes in the levels, composition and location of production, and in farm practices. Both economic and environmental benefits will likely flow from a more economically efficient use of resources in producing agricultural products. However, in countries and for those commodities with very low levels of support, there could be an expansion of agricultural production, as producers in these countries might be in a better position to exploit their comparative advantage in producing commodities in response to changes in market prices. This could place extra pressure on the environment. It should also be noted that shifts in consumer preferences (e.g. towards "environmentally friendly" agricultural and food products) expressed through markets, have also a strong influence on farming practices in some countries.

The linkages between farming activities and the provision of environmental benefits involve physical, chemical and biological processes, as well as sociocultural processes, which are often specific to local or regional situations. Many agricultural policy measures are still administered at national level on a commodity or farm basis, whereas the environmental effects of agriculture relate to resources, and are not only commodity or farm specific. These policies may thus create different environmental problems or benefits according to the local or regional conditions. As a result, the effects of agricultural policies on the environment involve non-linearities and uncertainties, which make quantification difficult. In this context the use of agri-environmental indicators can contribute to any assessment of the environmental beneficial or harmful impacts of policy measures. Efforts to establish a set of such indicators are currently under way in the OECD, and this Seminar can also contribute to throw further light on the measurement of environmental benefits.

2. Agricultural policy reform and environmental benefits

The effects on the environment resulting from the agricultural policy reform process will depend on the extent to which relative incentives facing farmers change and the initial level and means of support afforded farmers in different countries. The diversity of environmental outcomes will vary between farmers, regions and across countries. This could occur due to diverse effects on production mix (such as between livestock and crops), variations in the state of the natural resource base and different farm management practices. However, in the process of reform the level of agricultural output that would arise under more market-oriented conditions, and the farm practices that emerge, might not correspond to the environmental outcomes desired by society. This is where market prices may not reflect the full environmental costs of agricultural production, and may neither be adequate in determining the quantity of environmental benefits demanded by society, nor fully capture the harmful environmental effects of agricultural activities. Thus, there are some environmental effects that would need to be addressed by specific measures and approaches in addition to agricultural policy reform: market orientation is a necessary, but not necessarily a sufficient condition for this to occur.

While agricultural policy reform will alleviate some environmental problems, others will likely persist and new ones could be created, especially with respect to land use. Agricultural production is likely to expand in some countries or regions and contract in other areas, which could create pressure on the environment. To prevent degradation of the land used in or withdrawn from agriculture, following reform, complementary policy measures may be required. Marginal land that was brought under cultivation as a result of high support levels, could be withdrawn from production if it ceases to be profitable as farmland, even with lower-input "extensive" farming systems. Although in some areas, such land might soon revert to a more natural state and form part of a balanced ecosystem, some land could, if left to itself, degenerate. Proper environmental management of idled land resources over time could become important in the context of reform. However, it should be noted that some of the alleged problems from abandonment of land are not environmental, but relate to rural development.

Support reductions will free up not only land but also other resources used in agriculture, whose reallocation to other productive uses within and outside of agriculture may not follow a smooth path to the benefit of the environment. The farms that remain in the industry may not have the financial capacity to absorb the freed-up resources, and a move to less extensive production methods may not always be possible, as it often relies on the availability of off-farm income. Where resources are permanently withdrawn from agriculture, there can be a danger that they depreciate (human-made resources, such as farm roads and buildings, irrigation and drainage systems) or degenerate (natural resources). However, these effects can be mitigated where reductions in support levels are accompanied by efforts to ease the adjustment process in the sector. Such efforts could involve measures to increase labour mobility, stimulate rural development, and provide temporary compensation to farmers.

3.   Key policy issues associated with the provision of environmental benefits

The long-term management of environmental resources in the production of food, and industrial crops is the basis for ensuring the sustainability of the farming system, and provide other environmental benefits from agriculture. However, the cost of these resources has generally been assessed in terms of their contribution to food and fibre supply. Values of output are determined by prices determined in markets or by policies, and the cost of resources measured by farmers' expenses on inputs. Such a valuation, even where not distorted by market or policy failures, is often incomplete. This is because the agricultural activity itself also provides environmental benefits demanded by society, which are not accounted for in farmers decisions, or when the agricultural activity does not provide the level of environmental benefits demanded by society.

A move to more market orientation can ease the pressure on the environment and changes in agricultural production imply a change in the provision of environmental benefits. But where markets fail to take account of the value of environmental benefits from agriculture, policy makers may have to make explicit provision to account for environmental benefits and harmful effects which are part of the "by-products" of agricultural production. To do so it is necessary that such environmental benefits and harmful effects are identified and evaluated. In principle, the task of policy makers will be to encourage ways to provide the appropriate incentives and penalties to ensure that farmers, and others who control agricultural resources, use them in a way which reflects both their market value and their environmental costs and benefits. The goal is to make the choices facing farmers, and others, reflect the full social costs and benefits involved. The resulting distribution of resources would represent a higher attainable level of real income (including environmental values) for society given its available resources and technology.

In practice, there is no agreed definition of environmental benefits, nor methods by which they may be measured, which can unambiguously lead to the design of policies or approaches which provide the right incentives to farmers to encourage an appropriate level of provision. Nevertheless, many policies in place intend to compensate farmers for the costs of providing "environmental benefits". Government intervention which tries to cope with perceived environmental benefits, may distort the allocation of resources in ways which make society poorer in both market and environmental values. In this context, it is important that the Seminar discuss the possible roles of policy and markets, by adopting a pragmatic and forward-looking approach drawing on country case studies on the diverse experiences in OECD countries, and the other material prepared for the Seminar. The suggested central themes and questions outlined below are indicated to help focus the Seminar and identify the key policy issues associated with the provision of environmental benefits.

The nature and measurement of environmental benefits from agriculture

– *What is the nature of environmental benefits from agriculture and how can evaluation methods be developed to measure and monitor such benefits?*

- What are meant by "environmental benefits" from agriculture?

- What is the distinction between minimising harmful effects, avoidance of negative effects, and providing beneficial effects on the environment from agricultural activities?

- How could policy makers determine the quantity of environmental benefits that they should pay for? How can methods be developed to evaluate environmental benefits for use by policy makers?

– *What are the environmental benefits resulting from agricultural activities?*

- How do environmental benefits vary across OECD countries and farming systems, and to what extent are they linked with farming practices?

- To what extent are the environmental benefits identified with agriculture as such, compared with other land using activities?

Policy measures and practices and environmental benefits from agriculture in individual OECD countries

– *What is the experience of specific policy measures and practices to promote benefits from agriculture in OECD countries, and which policy instruments are used to achieve this?*

- What are the environmental benefits from agriculture identified or perceived in individual OECD countries?

- What are the policy approaches and programmes to encourage the provision of environmental benefits in OECD countries?

- What kind of instruments or mix of instruments are being used to enhance environmental benefits from agriculture, and how are the programmes assessed and monitored?

- To what extent are farmers own practices contributing to the provision of environmental benefits; and what has been the involvement in the policy process by farmers towards this provision?

The role of agriculture and agricultural policy in providing environmental benefits

– *To what extent can a combination of agricultural and environmental policy measures best achieve environmental benefits from agriculture?*

- How do different agricultural systems and policy measures affect environmental benefits from agriculture?

- How could agricultural policies be framed so that farmers' actions will maximise the beneficial effects on the environment and promote sustainable agriculture?

- What types of targeted environmental policies would need to complement agricultural policy reform to provide environmental benefits?

- What is the potential for governments in defining property rights and obligations, or using voluntary mechanisms, in order to provide these benefits with minimal financial transfers from governments?

- To what extent do the costs of providing environmentally beneficial agricultural practices justify financial compensation from governments?

- What might be the effect of agricultural trade liberalisation on the environmental benefits provided by agriculture?

4. Conclusions

The conclusions of the Seminar will draw on the discussion, and will address the key questions of relevance to policy makers:

– In the context of agricultural policy reform what is the role for other complementary policy measures to enhance environmental benefits and reduce environmental harm from agriculture in the OECD countries?

– To what extent and how can countries benefit from the experience of policies in other OECD countries in enhancing the provision of environmental benefits in agriculture?

– What is the role for governments in designing policies and approaches, and promoting market solutions to enhance the provision of environmental benefits with least production, trade and economic distortions?

SUMMARY OF THE SEMINAR

by
Anton D. Meister
Massey University, Palmerston North, New Zealand

Introduction

The relationship between agriculture and the environment is an issue that has steadily been receiving greater attention over the past years. While early discussions dealt very much with harmful aspects, more recent discussion has focused on beneficial aspects of this relationship. Thus, while on the one hand agricultural activities can lead to pollution of surface and ground water, destroy indigenous vegetation, drain wetland and cause visual pollution, on the other hand, sustainable farming systems can be associated with the maintenance of traditional landscapes, the preservation of natural habitats and biodiversity, improved water and soil management, and sustaining rural communities and cultures.

The need to enhance the positive and reduce the negative impacts of agricultural activities has become a major policy concern in all countries. The reasons for this are many and reflect a desire to maintain the scarce natural resource base, to enhance the health of the global environment, to increase the health and well-being of people, and to sustain rural communities and cultures. At the international level, countries have agreed to work towards this goal at the UNCED Earth Summit in Rio.

Recent agricultural policy reforms in OECD countries and the conclusion of the Uruguay Round have affected the structure and pattern of agriculture as well as its profitability, and this will continue. These changes are brought about by the lowering of levels of overall support to agriculture and the shift of policy from market price and other commodity linked support toward budgetary financed measures, as well as the lowering of trade barriers and the liberalisation of international trade.

The reforms taking place in the agricultural policy of OECD countries have consequences for the agriculture sector with resulting environmental impacts (both negative and positive). The direction of change in environmental impacts varies from country to country. However, to assure the enhancement of the positive impacts and a reduction of the negative ones, budgetary payments to farmers for undertaking specified activities to achieve these purposes have increasingly become part of the policy mix in many OECD countries.

This summary attempts to draw together the main threads that emerged from the papers presented and the discussion at the seminar. A key focus of the discussion dealt with the rationale for government support to providers of environmental benefits without causing production or trade distortion, on how to formulate policies using various instruments, and on policy monitoring and evaluation. The issues raised can be summarised as broadly dealing with:

- the nature and measurement of environmental benefits from agriculture;

- policy measures and practices in OECD countries to promote environmental benefits from agriculture;

- the role of agricultural and environmental policies, and market solutions to provide environmental benefits from agriculture, within an overall context of policy reform and a striving towards a more sustainable agriculture.

1.   Background concepts and framework

Environmental benefits from agriculture

When considering environmental benefits from agriculture it is important to clearly define what is meant by 'benefit'. For a given goal or objective, a 'benefit' is the result of some action that leads to an environmental outcome beyond a certain environmental level, while a 'harm' is the result of some action that leads to an environmental outcome below that level. Sandra Batie in her paper talks about environmental services as a continuum along a spectrum; while Daniel Bromley classifies the nature of those services as amenity, habitat, and ecological services. The way agriculture influences these services can be positive (benefit) or negative (harm). But it is important to recognise that enhancing benefits or reducing harm are not cost-free choices.

Amenity benefits, for example, are the large class of visual attributes of the rural countryside that make it pleasing (or unpleasant) to the visual sense. Given a specific level of amenity benefits, agricultural activities (intensification or extensification of farm practices, use or abandonment of land, monocultural or mixed farming practices, etc.) can enhance or diminish this level. If farmers provide more of the landscape than the expected level, they are providing amenity benefits, while if they provide less, then they might be accused of causing a harm. Hence it is clear that an environmental benefit (or harm) has to be viewed as relative to a reference level.

It is important to note that amenity benefits do not necessarily deal with natural environments (few of which exist anymore in Europe). We are talking here about manmade environments which society desires to retain or enhance. This is quite different from the amenity benefits desired say in Australia, New Zealand or North America where a preference is expressed for natural (or "wilderness") environments rather than man-made environments. This distinction shows that amenity benefits are very subjective and vary from country to country or even from region to region within a country.

Habitat benefits are to some extent less subjective in that we can measure the effects of a given activity on these habitats in terms of biodiversity, the ability to sustain wildlife or the presence of certain flower species. Agriculture, in that it uses nature to produce food and fibre, can have positive and negative impacts on the habitat provided by the natural environment. As Lewis Nowicki describes in his paper, agriculture over the ages has had a positive impact on the level of biodiversity in Europe but, in more recent times, this positive impact has turned to a negative impact. As with amenity benefits, society may have in mind a level of biodiversity it desires to maintain, and agricultural activities can provide more or less of these. How to define "more" or "less" requires us, firstly, to establish an environmental target. Secondly, a reference level has to be set, which defines the benchmark between avoidance of negative effects and the provision of positive ones. With society's definition of property rights, the reference level indicates the point where society's environmental objectives go beyond what can be imposed on farmers at their own costs.

Finally, there are ecological services. Daniel Bromley defines these services as the attributes of agriculture that affect, positively or negatively, ecological functions beyond the boundary of the farm. These services again can be negative, such as the typical externalities of agricultural activities that affect soil, water and air, or positive when agricultural activities reduce some of negative effects or minimise the negative effects of non-agricultural activities.

It became clear from the discussion that although all countries recognised that agriculture can provide some or all of the services mentioned above, the types and level of the associated benefits that are provided vary from country to country, and so does the value people assign to them. Hence, while in one country the value of man-made farming landscape is high, in others natural landscape (native flora and fauna with no anthropogenic intrusion) is preferred. It is therefore important to realise, from the start, that as the issues differ across countries, so will the policies and approaches.

Property rights, and environmental benefits and harms

As noted above, the distinction between benefits and harms becomes clearer when we define reference levels (papers by Daniel Bromley and George Hutchinson). Reference levels indicate what a society expects farmers to comply with at their own costs. Beyond the reference level society could pay farmers in their role as steward of the nation's natural resources. If agricultural activities provide more than the reference level, this can be seen as providing a benefit for which farmers, or the landowners, should be paid, just as is done for most other providers of services. When agricultural activities push the level of environmental services below the reference level, then society would be justified to call this a harm and require farmers to restore the reference level by mitigation or avoidance (the Polluter-Pays-Principle). But such a requirement depends on the concept of property rights which embody the rights and duties associated with the ownership of resources (particularly land in the case of agriculture): "Property rights define the boundaries of farmers' socially sanctioned activities and thus are central in the dialogue about agriculture and the environmental externalities. The boundaries of farmers' discretion are malleable, shifting as collective values and policies evolve over time" [Colby, (1995)].

Once reference levels and property rights have been clearly defined, it becomes easier to distinguish between providing environmental benefits and preventing environmental harm, as well as to embody rights and duties. It is clear that different OECD countries have different property rights structures with respect to land, which can change over time, and that there is no unique way to assign property rights structures and therefore assign reference levels. This has important implications for the environmental policies and the payments or non-payments associated with them in different OECD countries.

Environmental benefits from agriculture and the market

Another important aspect of the environmental benefits provided by agriculture is their character as being a joint product with commodity production. In some situations farmers themselves are able to capture the value of some of those environmental services (rural tourism, sales of local products etc.). But, this is not always the case, as the services have public good characteristics due to their incomplete excludability and their non-rivalness in consumption. Market-failure (the fact that the level of private benefits differs from social benefits) may lead to an under-supply of some of the environmental benefits which display the characteristics of a public good. This is, for example, the situation as portrayed in the Finnish case study where, "the greatest threat to rural landscapes are caused by discontinuing cultivation, depopulation of rural areas and closing of the open cultivated landscape." Hence the case for payments to assure that active farming remains and the landscape benefits are preserved. This contrasts with the New Zealand situation where there is little evidence of market failure in the provision of environmental goods by New Zealand agriculture. There, landscape amenities and *in-situ* preservation of biodiversity are by-products provided by an unsupported profitable agriculture which are provided regardless of payments from government.

If society desires a higher than the currently provided level of environmental benefits (or desires to assure maintenance of the current level) other institutional arrangements need to be relied upon to bring this about in cases where a situation of market failure is present. There are then good economic and social reasons for public policy to attempt to ensure an optimal supply of environmental benefits associated with agriculture. When, however, governments act to enhance the level of environmental benefits, it is important that the policies put into place to achieve this are consistent with the commitments made by OECD countries to agricultural policy reform and trade liberalisation.

Agricultural policy reform and environmental benefits

The history of most OECD countries is that governments have intervened in the agricultural sector through stimulating demand expansion and controlling supply. This has been undertaken through price and income support and associated trade measures, through risk reduction programmes or through subsidised inputs and budgetary transfers. The impact of this intervention on the environmental benefits provided by agriculture has been mixed, but on balance it can be said that high levels of support in agriculture have tended to intensify the adverse effects of agricultural production on environmental quality. This general statement has to be treated, however, with some care, as the influence of agricultural policy on the environment depends on farmer participation rates, land quality attributes and farming practices (paper by Sandra Batie).

Lewis Nowicki explains that some environmental benefits are possibly better assured in a context of liberalised trade as compared to a high support price policy. These are the kinds of benefits that most directly impact on human society and the sustainability of agriculture itself such as air, soil and water quality. A further conclusion he draws is that some environmental benefits may not be assured in such a context, and these have to do with the very substance of what makes Europe in many ways unique: the biodiversity and landscape values of cultural landscapes, maintained only through a living social fabric in rural areas, nourished by agricultural activity.

Policy reform, which reduces support and shifts from prices or inputs support measures, to measures decoupled from commodity production, may or may not bring about an increase in environmental benefits. A reduction in the intensity of farming will have positive benefits for the environment in terms of reductions in the use of fertilizer and pesticides, and pressure on the conversion of wetland. However, the reduced profitability of farming, in for example marginal land areas may reduce environmental benefits of the nature of landscape and habitat and may even increase some of the ecological harm (such as soil erosion and an increase in flood risk). Hence, the consequences of policy reform will be country specific. It was generally agreed that even though policy reform should enhance the potential for environmental improvement, it will not necessarily of itself achieve those improvements, and complementary environmental policies may need to be put into place.

A second reason for continued emphasis on agricultural policy reform is to improve coherence between agricultural and environmental policies. While on the one hand environmental policies strive for a cleaner environment and the preservation of habitat and maintenance of landscape, production oriented agricultural policies may offset the environmental policies by giving price signals that lead to increased fertilizer use, the expansion of monocultures to reap economies of size and also alter landscape to allow large-scale farming. Agricultural and environmental policies should be consistent, and co-ordinated across government departments.

There is, however, a final reason why agricultural policy reform is a necessary component of the process to enhance (or maintain) the environmental benefits of agriculture, and that is the need to get the policy framework right before environmental policies are devised. Environmental policies can be of various kinds, ranging from regulatory measures to financial incentives, and voluntary actions. When policies involve environmental payments or taxes, in a context of high levels of production-linked agricultural support, then these environmental payments or taxes need to be very high to be effective. This is because environmental payments (for services provided) may to some extent counteract the negative effects of agricultural support. In such cases farmers would be over paid for their joint product, and the result would be the achievement of environmental goals that are more costly to the economy than is necessary.

Agriculture and the environment: a partnership

There was a clear understanding that a partnership exists between agriculture and the environment. While agriculture and the environment have often been seen to involve trade-offs (more of the one means less of the other), this view has very much changed toward seeing the relationship as complementary and being able to provide mutual benefits to the partners and to society. But, as with all partnerships, to enhance the mutual benefits the following guidelines are needed:

- A good understanding of the relationship between the two partners: how does the one affect the other, and what are the values or benefits that each partner can contribute?

- Clear objectives and rules: what are the aims, and what are the rights and duties of each partner?

- Non-conflicting policies, and co-operation in policy design and implementation.

- The use of the precautionary principle whenever there is uncertainty in the relationship between agriculture and the environment.

- A holistic approach to overall management of agricultural and the environmental policies.

- To keep in mind the long-term consequences of current actions: some actions today have long-term and dynamic effects, and these should be considered before decisions are made.

Overall, while there was general consensus about the valuable contribution that agriculture makes to providing environmental benefits, there was also the realisation that there is a continuum of environmental effects which, depending on the reference level (which determines the rights and duties of farmers), can be classified as benefits or harms. This defines the policy environment within which policies are formulated to enhance the benefits and reduce the harms in an efficient and equitable way.

2.   Country case studies

The experiences from different OECD countries are wide ranging, dealing with a variety of cultures, agricultural systems, and environmental benefits and problems experienced. Therefore, no single policy approach may suit the particular situations faced by different countries. While this summary does not discuss the details of the various policy approaches, it attempts to show that a lot of progress has been made in encouraging the environmental benefits of agriculture. Some of the key issues raised in the discussion of country experiences, in formulating and implementing policies are dealt with in the following paragraphs.

Current policies and instruments

Agri-Environmental Regulation 2078/92 has provided the overriding framework within which **European Union**'s programmes are shaped in each Member State. Although implementation of the programmes is obligatory at the Member State level, farmers may choose whether to continue their normal farming or to join - usually by contract - an agri-environmental programme. The measures supported fall into four broad categories: low-intensity farming systems; landscape; set-aside and maintenance of abandoned land; and training and demonstration projects. Thus, for example, the first category requires:

– stopping the use of chemical fertilizer and pesticides, and using organic farming methods;

– reducing the number of cattle and sheep per hectare of forage land;

– converting arable land into extensively used grassland;

– setting aside arable land for environmental purposes for at least 20 years;

– maintaining land withdrawn from agriculture and forestry.

Examples of these were provided in the case studies of countries in the European Union. In **Germany**, the Lander programmes encourage farmers to reduce fertilizer and pesticide use and payments are made for income lost. As these payments are based on average income lost (plus a lump sum incentive) the creation of rents cannot be avoided between and within regions in Germany. In **Austria**, a major change towards organic farming (see the first condition above) has been encouraged with the aid of payments to achieve a more sustainable agriculture with associated impacts on the health of the soil, water, air, and food products, as well as maintenance (or even enhancement) of biodiversity. In the discussion on the Austrian case study, while it was accepted that organic farming by definition is more environmentally benign than capital intensive farming, it was recognised that more scientific evidence is needed to demonstrate the superiority of organic farming over conventional farming.

A typical situation found in several OECD countries is that agriculture (often small-scale) is disappearing from the more marginal land areas (as for example in Norway, Finland, Greece, Switzerland). With the disappearance of agriculture, some of the environmental benefits may also disappear (especially those associated with landscape and cultural values). In these situations the agricultural landscape (the agricultural activity with the associated buildings and crops) is the environmental benefit and therefore it is deemed necessary to maintain (or even enhance) agricultural activities. In the case of **Norway** an Acreage and Cultural Landscape Scheme is in place to maintain land in agriculture and to maintain the landscape and cultural heritage or, as stated in the case study, "preserving-by-using". The grants provided by the Scheme are not linked to production and actually encourage less intensive farming. The payment applies to all arable land, sown grassland and fertilized pasture. Farmers have to apply for the payment, and need to satisfy certain obligations.

The situation in all of these countries is that the environmental benefits are at risk of disappearing if agriculture is no longer practised. In **Switzerland** the overall programme consists of research, training and extension to heighten farmer awareness of environmental problems and possible solutions; financial incentives; and authority to issue mandatory directives. As is pointed out in the case study, these measures serve more than one objective: social (maintain small farmers and villages in the rural area), food security (supply the population with food), environmental (safeguarding the countryside, building and pathways, and preserving the biological diversity of rural areas), sustain resources and minimise wastes. The aim is not to just have farmers in the region, but to have profitable farmers. While policy measures may be designed to achieve one specific objective, they may also address some of the other objectives outlined above, reflecting the integrated nature of agriculture. As these objectives can not really be considered in isolation, and policy measures may have conflicting effects, it is important to ensure that they are complementary and well targeted.

Although most of the programmes discussed in the country case studies are optional and use payments to try to change behaviour within a legislation framework, there are situations where voluntary action is possible without regulations or incentive payments. Sometimes this action is precipitated out of fear that legislation may follow if nothing is done, or it results from consumer pressure, or it is simply the desire to fulfil the stewardship role associated with land and resource ownership. An interesting example of such a situation was presented in the **United States** case study on biodiversity. In this case, private sector initiative by a landowner-driven organisation led to an attempt to implement ecosystem management on nearly one million acres of virtually unfragmented open space in Arizona and New Mexico. It is a private initiative by ranchers that brought together local, state and federal government personnel to develop co-ordinated solutions to the environmental concerns identified. To some extent we see this also happening in the shape of Landcare groups in **Australia** and **New Zealand**.

In the **United States** case study on protecting biological diversity, the instrument of transferable development rights is discussed as a tool to stop the conversion of agricultural land for urbanisation by providing a buffer zone to protect a major wetland. In the US case study on the maintaining and enhancing the environmental benefits of wetlands, farmers have been encouraged to sell easements on enrolled wetlands in the Wetlands Reserve Program and to carry out a wetlands restoration plan. In return, they receive payments for the easement and funding of the wetland restoration costs. Also associated with some of those programmes are initial steps to introduce wetland mitigation rights which are transferable and in that way compensate farmers for wetland restoration and preservation costs [Fulton, (1996)].

One way to achieve environmental objectives is to carefully target the areas that need to be managed in a special way to achieve clearly specified environmental objectives. An example of this approach is the management areas scheme explained in a country case study by the **Netherlands**, and the environmentally sensitive areas scheme (ESA) in the **United Kingdom**. After the management objectives have clearly been specified, prescriptions are formulated and combined into a package of management measures. A management agreement with farmers allows for compensation, based on the principle that "farmers should not suffer a decline in income in comparison with a farmer who operates under similar conditions but does not conclude a management agreement (reference situation)". In the UK scheme, a series of Codes of Good Agricultural Practice (GAP) are prescribed, which provide a benchmark against which both regulatory and voluntary incentive policies are determined. Payments are only made for activities which go beyond GAP as defined by the Codes. In both cases compensation is based on income foregone, to which additional payments encourage farmers to enter the scheme.

In some countries, land is withdrawn from agriculture and managed as a nature reserve by either a nature conservation authority or by private landowners. Examples are the nature reserves in the **Netherlands** where farmers are paid compensation for the land withdrawn and, if they manage the reserve, a management fee. An alternative arrangement would be a trust or easement on the land (this may be part of a farm) which withdraws it from agriculture and is managed as a nature reserve. This is also found in **New Zealand** (the Queen Elizabeth II trust), where no compensation is paid for the land withdrawn, but some financial help is provided for fencing off the land from the rest of the property.

The **United States** case study on protecting biological diversity also highlights the role of partnership programmes. These programmes demonstrate how private landowners, local, state and government agencies and non-governmental agencies can work together in using resources for protection strategies that are larger than the sum of each of the individual parts. The cases presented illustrate how The Nature Conservancy has served as a catalyst to bring together government departments, non-governmental organisations, local and central government as well as private land owners and business.

An approach helping farmers to devise a sustainable plan for their farms has been used in **France**. The overall aim is to establish systems that are economically viable, safeguard environmental standards and contribute to local development. For each of the farms in the programme, an analysis of the local situation is performed which is followed by an agricultural and environmental assessment of the farm. Then a scenario is built that incorporates the strengths and weaknesses of individual regions and farms, and finally a contractual agreement (five year contract) is signed between the farmer, the government and other partners. Under the contract the farmer can draw on agricultural and environmental assistance available in the region, and will receive training, advice and back-up services, together with a special grant. While this programme will achieve environmental benefits from agriculture, the linkages are not always clear between payments and environmental outcomes.

In the discussion it was often mentioned that 'We know so much but tell so little', which simply reflects the need to go out to explain and demonstrate to farmers. The **Korea** case study dealing with the problem of soil erosion and soil degradation is based on extension, giving farmers advice on optimal fertilizer use and agricultural practices, as well as getting them to use soil surveys on a regular basis. In the discussion it was stated that "education is a must; you can't just pass laws and regulations and expect people to follow." Farmers often want to do what is right, but they often do not know what is right. Through extension work and by involving farmers in demonstration programmes - doing things with them rather than to them - much can be achieved.

Overall assessment of policy experiences

Through the discussion of the country case studies, the seminar participants obtained a much better appreciation of the wide range of environmental services provided by agriculture, such as for example, the natural bush and shrub preserved by New Zealand and Australian farmers, the quality of rivers, lakes and air in most countries, the cultural landscape of the Greek islands, the amenity benefits of mountain and hill farming in for example France and Switzerland, the moors and heather included in some of the ESAs in the UK, the wetlands in the US, the ecological services of rice paddies in Japan and Korea, and so on. All of these services can provide benefits to society ranging from use values (for example recreation and aesthetics) to ecological values and option and bequest values.

The country case examples also show that there is a range of policy approaches and no one policy that serves all situations, or that the continuation of agriculture is always the best and only option. Sometimes it may be best, to achieve long-term preservation of the environmental benefits, to acquire land or easements and discontinue agricultural production (often the original cause of the environmental benefits being under threat). Further, the approaches show that it is not always government regulations combined with compensation payments that are required to achieve objectives. Voluntary approaches, co-operation between private individuals and public agencies and industry, research and extension and education can all go a long way towards achieving the objective of maximising environmental benefits from agriculture[1]. While in some cases government funds are provided for the work of those groups, in other cases they are purely voluntary and receive no funding at all, but still bring about significant changes by moving agricultural activities to more sustainable outcomes.

Where environmental benefits are a joint product with agricultural output, it is clear that agriculture should provide these benefits. Nevertheless, this was an area of some controversy, since what seemed to be implied by some countries was that only by reversing the declining profitability of specific forms of agriculture could the loss of valuable landscapes and semi-natural areas be avoided. The controversy was not so much about agriculture being the provider, as about what was needed for agriculture to be able to provide the benefits. The question raised was whether agriculture necessarily had to be maintained (through agricultural support) to achieve the desired benefits? Were there alternatives, such a paying farmers for preserving the environmental services, without trying to maintain the profitability of agricultural production itself?

While some participants claimed that paying to keep agriculture profitable represented simply a way to keep uncompetitive farming systems which could lead to production and trade distortion, others commented that this simply reflected a correction of market failure and that this was not therefore creating distortions. Then, where the environmental objectives of a programme or policy are clearly identified, and payments (or compensation) are specifically made for the achievement of those specific objectives, this can be seen as correcting a market failure. However, when the objectives are not clearly spelt out, and payments are made to agriculture to make it more profitable to keep people on the land, with the presumption that this will lead, (without clear proof of a linkage between profitable agriculture and environmental benefits) to an increase in environmental benefits, then this situation could be seen as production and possibly also trade distorting.

But the question was asked at the seminar, "is supporting the agricultural system the only way to bring about the desired environmental benefits?" In some cases for example there may be alternative and cheaper ways to achieve an increase in biodiversity, prevent soil erosion, or to protect some ecological values. The point here is that even though in most cases the maintenance of agricultural systems is the most practical and efficient way to bring about the environmental benefits, it should not

---

[1]   For additional ideas on possible policy alternatives, see two recent OECD publications, 1) Amenities for Rural Development. Policy Examples [OECD, 1996a] and, 2) Saving Biological Diversity. Economic Incentives [OECD, 1996b].

always be presumed that this is the only way. This latter question would need to be addressed in each particular situation, and not just accepted as such.

The issue of production and trade distortions was raised in that agricultural production should not be the prerequisite for receipt of incentive measures: direct payments to farmers who provide (or agree to maintain) certain desired levels of particular agri-environmental attributes can affect farmers' production decisions and agriculture's productive capacity. However, other participants stated that the provision of environmental services is in many cases unavoidably linked to well adjusted agricultural activities. The risk of distortions can be minimised if incentive measures are transparent, have measurable performance indicators, and are clearly tied to environmental objectives. Moreover, the cost effective means of achieving the desired level of environmental attributes could in some cases be obtained with other than through agricultural production, and this should always be investigated.

While the above country examples have only briefly touched on some of the approaches presented, they do give a flavour of the range of tools and policies available. What the examples show is that social, economic, rural development and environmental objectives are often mixed together, which lessens the transparency of the programmes, and incentive payments might be seen as much to keep farming as to provide environmental benefits. Because of the wide variety and range of environmental benefits from agriculture it is important that the objectives of environmental policies and programmes are clearly spelt out in order to enhance their transparency.

Another reason for clearly spelling out environmental objectives is to help establish indicators for monitoring and evaluation. It is only when one knows what is to be achieved that achievement and success can clearly be measured. However, when agricultural policy measures are used to achieve environmental benefits, it is important to be clear as to what are the environmental objectives and what are the social and rural development objectives. Of course, the clear separation of activities pursuing the various objectives may not always be possible since, especially in the European situation, agriculture and environment are closely interlinked when talking about landscape and cultural values.

For example, as shown in a number of case studies, it is difficult to separate the contributions of the agricultural, social and environmental policies to maintain extensive farming systems over a large area of mountainous country. But, even here, the aimed for environmental benefits should be clearly spelled out, so as to avoid the appearance that environmental payments are simply substitutes for price support payments. In the Greek example this was done by identifying the environmental benefits that would be lost if land was abandoned or not farmed: protection against soil erosion, management of water resources, and management of habitats, agricultural landscapes and cultural heritage. Hence, while in this case study the objectives are clear, the related question has to be asked "are the best means being applied to achieve them?" This is something that will be discussed in the next section.

3.   Policies for the future

Rationale for environmental payments

The need for payment or non-payment for the provision of environmental benefits depends on reference levels and property rights. If, with regard to what society expects from farmers (the duties associated with rights in land use), they do less than the reference level (cause pollution - as defined by the authorities - that is affecting the environment beyond what they are authorised to do on the basis of clearly defined rights), then the Polluter-Pays-Principle (PPP) agreed by OECD countries requires that the polluter has to bear the pollution abatement costs. Or, in other words, "in some cases some might argue that farmers are failing to meet their perceived responsibilities implicit in the social contract that society holds with landholders as to the rights of land ownership. In these cases PPP would apply." [OECD, (1966a)]. No matter whether a policy measure has to deal with non-point source pollution or with reduction in landscape, amenity or habitat benefits, attempts should be made to apply the principle. There seem to be a much greater willingness to implement the PPP principle when dealing with industry than when dealing with agriculture. Equity and economic rationality demand consistency in dealing with different sectors in the economy.

In the converse situation, when farmers through their activities provide environmental services desired by society, beyond the reference level, then a benefit is provided and the beneficiaries should pay. Since many of these benefits come in the nature of a public good (i.e. it is difficult to collect payment from beneficiaries since they cannot be excluded if they haven't paid), then society should pay for the services and recoup the money through taxation or other means.

The question then revolves around the definition of the reference level. In the country case studies presented, several countries have moved towards defining "codes of good agricultural practice". This is a step in the direction of clearly defining reference levels, and what society sees as the duties of landholders as stewards of the land. Such a code would provide a good benchmark to help define benefits and harms resulting from agricultural activities. There is no easy and unique way of identifying this level, and the way rights and duties are specified depends on society's view of property rights. Rights structures vary from country to country and reflect cultural and political values, and it is each country's prerogative to make decisions as to how to frame those structures. In doing so, however, it is important that the wider consequences of such decisions are considered, especially to ensure their compatibility with the commitment to agricultural policy reform.

One consequence is the long-term incentive effect. As Daniel Bromley states, "Farmers will argue that they have a 'right' to do something and will insist that a denial of the 'right' calls for compensation". The consequence here is that once compensation is forthcoming, the "rights claim" of farmers becomes politically reinforced. Governments should avoid those policy actions that may bestow further rights (entitlements).

There was a general understanding that payments to provide environmental benefits should not be linked to commodity outputs, but rather targeted to areas put under certain environmental constraints, to environmentally sound production techniques or, where possible, to environmental outputs. However, this may not always be possible, as environmental benefits are often by-products of commodity production, and environmental outputs may be too difficult to measure to link payments to them. This was discussed, for example, in the German case studies, where reasons are given why the linkage may not often be possible. But, that does not mean that attempts should not be made to implement it as often as possible. It was also recognised that environmental payments should be directly related to the costs of achieving environmental outputs or to income foregone, and should be balanced against any environmental harm that agricultural activities incur. This again raises the issue of transparency and accountability, as well as avoiding supporting agricultural activities on the one hand that harm the environment and then on the other hand paying for those same activities to minimise those impacts.

Improving policy formulation

With regard to improving policy formulation, a number of points raised in the discussion should be noted. One is that the long-term consequences and the dynamic effects of policies should be considered. For example, what are the implications for land prices, production patterns and trade, of payments that still to some extent keep resources in agricultural activities and encourage production? The German case study, for example, discussed the creation of economic rent from payments that were given to farmers irrespective of their different situations.

A second point is that in devising policies for the provision of environmental benefits, the aim should be to ensure the long term preservation of such benefits. Benefit preservation that is mainly dependent on budgetary payments may be at risk, as Governments and economic conditions change and so could the willingness to continue paying. Therefore, where possible, permanent solutions should be looked for, such as land acquisition, permanent easements, and changed management systems. For example, in the Japanese case study, the preservation of paddy rice fields to prevent soil erosion and achieve better water management for flood control, may be the best, short-term solution to achieve these benefits, but, a longer-term solution could be to consider the afforestation of carefully targeted land. Looking at policy purely in terms of environmental benefits, afforestation may well turn out to be the least-cost solution [OECD 1995]. However, if at the same time other social benefits are to be fulfilled, then keeping agriculture together with environmental payments may be the best solution, but this should be clearly stated and payments should clearly identify the objectives to be achieved. Possibly a mixture of afforestation and farming in hill country areas could be the best approach.

Thirdly, whenever possible, policies should be well tailored and targeted. Several of the case studies discussed programmes that represented 'blanket' policies. In certain situations this may be the only way to achieve the environmental objectives, but in others, this does not appear to represent a least cost approach. As Sandra Batie remarked "if seven out of two hundred farmers cause a problem, you should not include all two hundred in your programme but target the seven".

Tailoring and targeting therefore needs to receive more attention. One way to achieve this is by working much more at the local and regional level (the subsidiarity principle). This, besides allowing the tailoring of programmes closer to local needs and problems, also provides a better opportunity for greater involvement of people and farmers at the local level.

In improving policy formulation a number of questions arise: What is at risk? What will happen if for example agricultural activities where not supported in a particular area? What would the environmental impacts be? Once this has been done, programmes and policies can then be carefully formulated to avoid the loss of the particular environmental benefits at risk. The fact that land may be abandoned doesn't necessarily imply that biodiversity or other environmental benefits will be lost. This may happen but - for reasons of transparency and success of programme - the effects should be evaluated.

Measurement of achievement

Successful policies are those which achieve objectives at least cost. While some of the country case studies stated that good results had been achieved, not much specific evidence was provided of what had actually been achieved. This was pointed to from time to time by representatives of the Royal Society for the Protection of Birds (RSPB) and World Wide Fund for Nature (WWF).

Monitoring and evaluation was an issue running through the discussion. In many programmes, large amounts of money are spent to achieve environmental objectives and societies will want to see environmental results and hence the need for monitoring and evaluation. Monitoring and evaluation is necessary to see if objectives are achieved and to see that the gains to society are worth the costs, as well as to learn from successes and mistakes so as to be able to fine-tune or reformulate existing and new policy measures.

A related issue is the level of environmental benefits. Are more environmental benefits always better? There are many difficulties in trying to determine the welfare maximising level of environmental benefits. The determination of maximum welfare raises many questions, such as whose point of view should be taken, what benefits to consider, how about sustainability requirements and so on. The benefits themselves may have use, option, preservation and bequest values. To measure all of these and place them in the right sustainability context presents many difficulties of identification, measurement and evaluation, some of which are explained by George Hutchinson. Daniel Bromley states that it is not yet possible to present a model to calculate the optimal level, and we should probably leave it at that and accept that the optimal level will be country specific, being a function of equity, efficiency, the current state of the country's development, its culture and its values, etc. The optimal level will be unique and will be a movable target that changes over time as society's values and technologies change.

George Hutchinson, in his paper, introduced some of the theoretical and practical issues associated with physical monitoring and monetary valuation. While, as he explained, it would be valuable to have physical indicators of change in environmental states (for example landscapes or ecosystems) this is not always easy and it has been found that sometimes it is more practical to develop indicators of trends in driving forces or pressures, such as changes in chemical inputs, farm management practices and percentage of land area held in protected categories. While such a "Driving Force-State-Response Framework" would be informative and useful for simulation modelling, the need for a physical inventory remains. Improved technology is making this more feasible as shown, for example, by the Natural Resource Inventory in the USA which monitors 300 000 sites for landscape, wetland, agricultural practices, soil erosion, and layers of wildlife habitat.

One of the reasons for not having clear results as yet from monitoring programmes is that the agri-environmental programmes have only started recently, and it is too early to see results. However, some countries - the US, UK, the Netherlands and France for example - reported on findings from monitoring programmes, and they reported successes as a result of the programmes implemented. Another reason for not having too many results from monitoring as yet, is the difficulty some countries are having determining the indicators that should be used for monitoring. Indicators for water or air quality are not so difficult to design, but when it comes to landscape and habitat preservation it becomes more difficult. Even when talking about wetlands, "hectares of wetland" are limited as an indicator, since the biological value of wetlands differs so much. The Norwegian case study provided some information on indicators used to measure landscapes. The ones chosen are: distribution and size of isolated cultivated plots; distribution and size of depicted uncultivated areas/patches with emphasis on their significance to natural heritage; spatial resolution or scale of landscape, and agricultural structures.

With regard to monetary valuation, this has not been applied in many countries. The UK case study reported on a contingent valuation method approach that showed that the benefits of the particular programme outweighed that cost. Still, as George Hutchinson explained in his paper, valuation is not easy. Experience to-date should give us some healthy scepticism of the results obtained from non-market valuation techniques when valuing things like landscapes and habitat preservation. There is need here for more work to develop these tools so that they can be used to help guiding policy choices.

Overall, the discussion clearly demonstrated that there is a need for better indicators, for establishing clear criteria in the design and implementation of policies and for a continuing development of monitoring and evaluation techniques. As discussed above, accountability of the money expended in the programmes and the need for transparency demands it. There is clearly a case here for further work by all countries to develop indicators for particular countries and ones that can be used for intercountry comparison. Especially with regard to the latter, the seminar saw a role here for the OECD to guide this work.

4.   Concluding comments

The seminar generated a very good discussion of the issues associated with agriculture and environment, and provided much information on how different countries have dealt with the issues. The seminar did however, reveal differences in OECD country's philosophies and approaches, as was to be expected given the wide diversity in cultures, agricultural systems, levels of development and environmental problems. The discussion showed the need to further advance the OECD's work on developing environmental indicators for agriculture.

The seminar was also a learning experience. It has brought about a better appreciation of the different situations faced by different countries. The contrast between the situations of the European countries, Japan and Korea, and Oceania and North America was especially great. While land abandonment in Australia and New Zealand may not necessarily lead to a reduction in biodiversity as the land reverts to bush and shrubland, in Europe such a situation often risks leading to a loss in biodiversity caused by the removal of the agricultural systems which provide the current biodiversity. Similarly, the attitude of different countries towards rights and duties held by resource owners also varies significantly, with the result that in some countries changes can be implemented without compensation, while in others payments need to be associated with nearly any required change.

Despite those differences, there was a general understanding on how to ensure that agriculture continues to enhance the environmental services that it provides. The understanding basically refers to the aim of the policy setting framework, which is to enhance the environmental benefits and reduce the environmental harm. The discussion showed that while it was recognised that all countries formulate policies that fit in with their cultural and social values, policy formulation should give due consideration to shape policies in a framework of continuing agricultural policy reform, recognising the UR agreement on trade liberalisation, and thus reduce, as much as possible, production and trade distortions.

5.   Bibliography

COLBY, B.G. (1995), "Bargaining Over Agricultural Property Rights", *American Journal of Agricultural Economics.* Vol. 77, p. 1186-1191.
FULTON, W. (1996), "The Big Green Bazaar", *Governing*, June, p. 39-42.
OECD (1996a), Amenities for Rural Development. Policy Examples, Paris.
OECD (1996b), Saving Biological Diversity. Economic Incentives, Paris.
OECD (1995), Forestry, Agriculture and the Environment. Paris.

ENVIRONMENTAL BENEFITS OF AGRICULTURE: CONCEPTS

by
Daniel W. Bromley
University of Wisconsin, Madison, USA

Five questions must be addressed in the quest for clarity concerning the environmental benefits from agriculture. These concern:

— What are "environmental benefits"?

— What is the difference between the provision of environmental benefits and the prevention of environmental damages?

— What is the role of property regimes in the distinction between creating benefits and preventing damages?

— To what extent are environmental benefits unaccounted for in the cost and revenue calculations of farmers? and

— What are the policy implications of the "polluter-pays principle"?

The task is complicated because the very idea of a "benefit" is socially constructed. That is, a wetland may be seen as providing a "benefit" by one party (say an ecologist), and as the source of "harm" by another (say a farmer). That is, one who cares about waterfowl habitat will regard a wetland as a beneficial attribute, and a farmer seeking more cultivable land will regard it as a harmful interference with other desired objectives. In Asian agriculture, a rice paddy is beneficial to the farmer, but may be not be so regarded by an urban resident or a naturalist[2]. Or, the burning of crop stubble may allow a farmer to control pests, but may cause damages to those harmed by the smoke. So the same physical condition - or act - can be beneficial or damaging depending upon whom we ask. For that reason, I will use a more general approach in which I talk of the environmental implications of agriculture - some of which will be beneficial, and some of which will be harmful. But this begs the question of what is "beneficial" and what is "harmful".

The answer, which unfortunately begs yet another question, is that a "benefit" is something that moves us closer to some goal or objective, while a "cost" is something that moves us away from that goal or objective. The issue, we see, comes down to a collective determination of goals and objectives.

---

[2]   It must also be noted that rice paddies provide important buffering for monsoon rains, thereby reducing flood damages elsewhere.

The problem with this realisation about benefits and costs is that it affirms that policy formulation - and policy analysis - has no anchor in absolute truth. In a world that celebrates hard-edged analysis and great precision, it is perhaps disconcerting to realise that policy analysis depends upon some idea of the "public interest" that has no solid analytical core. Despite the well-developed analytics of welfare economics, public policy can never completely escape its linguistic and conceptual roots - politics. There simply is no policy without politics for the simple reason that policy concerns: collective intentions; collective rules; and collective enforcement of new behaviours such that the collective intentions will be realised. Policy is relatively easy in a dictatorship. In a democracy, policy is a constant struggle among competing visions of the public interest. This means that different interests - with differing and conflicting visions of the public good - will have different ideas about benefits and costs.

Those differences can be clarified if we are careful in how we describe and characterise the environmental implications from agriculture. But we cannot assume that this clarification will resolve the conflicts in policy. Indeed it often happens that clarification simply reinforces the divergent positions of the protagonists. In the interest of some conceptual clarification, I now turn to the problem of how we might classify or categorise the environmental implications of agriculture.

1.   Classifying environmental implications

I suggest that there are three general classes of environmental implications from agriculture: (i) amenity implications; (ii) habitat implications; and (iii) ecological implications. While these cannot possibly represent mutually exclusive categories, considering them in this light allows us to focus on important policy dimensions that might otherwise be obscured.

Amenity implications

By amenity implications I mean the large class of visual attributes of the rural countryside that make it pleasing (or unpleasant) to the visual senses. The pleasing rural landscape of northern Europe with its immaculate farms comes to mind here. The sweeping vistas of the Ile de France, the orderly rice paddies of Japan, and the hedged paddocks of southern England comprise what I mean by the amenity implications of agriculture. The rural landscape is both created and managed by agriculture, and this rural character is important in its own right. This serves to remind us that agriculture produces both commodities and amenities [Bromley and Hodge, 1990].

The policy problem arises because there is a market for agricultural commodities, but markets are missing for the amenity aspects of agriculture. This means that changes in the rural landscape dictated by the imperatives of agricultural production may not always be appreciated by those who see agriculture as providing both commodities and amenities. Consider the problem of changes in the agricultural landscape that alter the collective perception of its amenity values. If we imagine a continuum as in Figure 1, we can suppose that the current situation is defined by L*. Here there might be some dispute about the direction of change from the status quo (L*). If policy pushes farmers to provide a "more desired" landscape then they might believe they are "providing" amenity benefits and should be compensated accordingly. On the other hand, if farmers seek to provide a less desirable landscape - to reduce the amenity aspects of the countryside - then they might be accused of causing "harm" and perhaps should be charged accordingly.

Figure 1. Amenity implications of agriculture

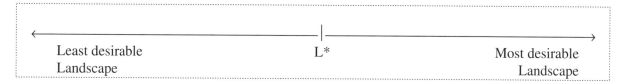

We see here the problem of the arbitrary nature of the *status quo ante*. Notice that L* is simply the momentary assessment of the amenity attributes of the rural landscape. This level of amenities becomes the norm against which a change in policy will be evaluated. Assume that urban interests begin to advocate a different landscape in rural areas. We might think of this as L_U. Farmers, in reaction to this might insist that L_F is really the appropriate level of rural amenities and they are already providing L*. In this setting, the distance $L_F - L_U$ becomes the bargaining space for this particular policy dispute. Urban politicians will advocate L_U while rural politicians will be likely to advocate L_F. We might regard these two points at the extremes of the bargaining space as the reference points for the two positions; it is to these two points that the political process will refer to resolve its disagreements (Figure 2).

Figure 2. The policy space for amenities

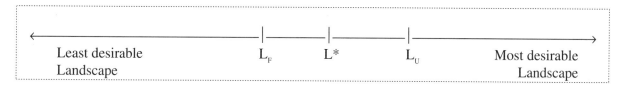

The essence of the policy debate over amenity values from agriculture is that there is no "right" or "correct" level of rural amenity. There are landscapes that are more appealing than others, but there are few precise rules that indicate the correct landscape. Hence the amenity implications of agriculture are probably susceptible to more serious disagreement in the policy process than are the implications of agriculture for habitat and ecological processes.

Habitat implications

By habitat implications I mean those attributes of the agricultural landscape that provide space and sustenance for plants and animals that are not part of the agricultural enterprise. As with the amenity dimension of the agricultural landscape, this aspect tends to focus on the land and water resources directly associated with the land in farms. In European agriculture, habitats are provided for birds and small mammals, and for native plant species. In Asian agriculture, the habitat component serves fish, small mammals, and native plants. In North America, agricultural habitats support wild game, waterfowl, and a range of native plants.

Unlike the amenity values discussed previously, the habitat implications of agriculture entail more certitude regarding the reference points for policy. Waterfowl require certain minimum areas for nesting. Wildlife require certain feeding conditions and cover. Fish require water of a certain temperature and purity. Wildflowers and songbirds also have certain ecological circumstance necessary for their survival. When agricultural practices are undertaken in a manner to assure that these minimum circumstances are met, then some would argue that farmers are "providing" habitat benefits. When farmers leave swaths of natural vegetation this would be an example of a beneficial habitat implication of agriculture. On the other hand, some might argue that in the absence of agriculture there would be even more of these circumstances and hence the very presence of agriculture has diminished the habitat component of rural areas. When locally unique and valuable habitats are destroyed by farmers then this is an example of damages from agriculture. As with the amenity implications of agriculture, problems arise because the habitat implications of agriculture are characterised by missing markets; there is a market for agricultural products but there are no markets for the habitat aspects of farming.

Wetlands currently constitute a controversial habitat conflict in North America. We might imagine a situation as depicted in Figure 3 in which H* represents the experts' views regarding an absolute minimum level of wetlands in a particular agricultural region, while H_S represents the *status quo ante* level of wetlands. To the right of H* we see the two reference points - H_F for the farming community and H_N for the naturalists who advocate far more wetlands than currently exist (H_S). While farmers do not necessarily seek to push total wetland area down to the absolute minimum (H*), their preferences are to have less wetland area than at present, while naturalists favour larger areas devoted to wetlands.

Figure 3. Habitat implications of agriculture

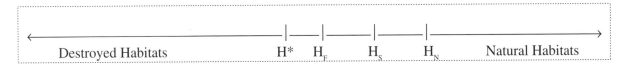

As with amenities, the policy response to this situation will differ. Farmers will fail to understand why they should be prevented from moving from H_S toward H_F. Indeed, we sometimes see pressures for financial compensation of American farmers for the lost income from holding them to H_S rather than allowing them to modify natural habitats to H_F. And of course the naturalists seek to have wetlands restored so that movement in the direction of H_N is achieved. Farmers will suggest that by being restrained to the *status quo ante* they are being made to provide habitat benefits for which compensation should be forthcoming. Naturalists will insist that agriculture has, in fact, destroyed the vast majority of wetlands present at the time of European colonisation of North America and that censure - not compensation - is the proper public response. There must be similar habitat debates in other OECD countries. How does one sort out these disagreements? As will be discussed below, the answer turns on the presumptive property rights situation.

Ecological process implications

By ecological process implications I mean those attributes of agriculture that affect, positively or negatively, ecological functions beyond the boundary of the farm[3]. One example is present when agricultural chemicals contaminate downstream rivers and lakes. Or when soil erosion clogs downstream waterways. On the other hand, farmers can undertake land-use practices that restore and enhance ecological processes. A wetland can act as a filter and a sink for certain agricultural

---

[3]   Notice that I regard the ecological processes as those beyond the boundary of the farm. This is an analytical convenience only and need not obscure the essential issues under discussion.

chemicals, thereby enhancing downstream ecological processes. Improved soil management practices and contouring reduce erosion and thus enhance downstream water quality. Finally, we must recognise land conservation as an important ecological implication. I have earlier talked of erosion control but it is important to stress that paddy field agriculture in monsoon Asia plays an essential role in stabilizing entire mountain sides. We might imagine serious ecological disasters in the absence of the water-control attributes of paddy agriculture.

We can use the familiar figure from above to assess the policy problems associated with the ecological aspects of agriculture. For simplicity, assume that nitrate contamination of groundwater is the problem to be addressed. Let N^* represent the upper threshold of nitrate concentrations beyond which medical experts warn that widespread and serious health effects will be prevalent in the general population[4]. Assume that the *status quo ante* level of nitrate concentration is N_S while farmers believe that nitrates in groundwater will not become a problem until the concentration level reaches N_F. Ecologists, on the other hand, might be expected to advocate a concentration level much closer to N_E (Figure 4).

We see once again that perceptions of what is correct will differ markedly across the various interest groups. Farmers can be expected to advocate a nitrate level that is somewhat higher than the status quo ante, but would certainly resist efforts to reduce it below the current level (N_S). They might not feel comfortable advocating moving too close to N^*, but they would probably stick hard to N_S if not necessarily N_F.

Figure 4. Ecological implications of agriculture

These three categories of environmental implications from agriculture will be used to develop ideas about the distinction between providing benefits and preventing harm, about the role of property regimes in this distinction between benefits and harm, about the publicness of environmental implications from agriculture, and about the implications of the polluter-pays principle.

2. Providing benefits or preventing harm?

As the previous discussion makes clear, much difficulty arises from a general confusion in the matter of whether particular agricultural activities and practices provide benefits or prevent harm. If a farmer decides to drain a wetland to grow crops then some would say that this act causes harm. Using the above classification scheme, they would insist that amenity aspects of the rural landscape would be diminished, important wildlife habitat would be destroyed, and ecological processes would likely be undermined. The farmer, on the other hand, might be inclined to suggest that if he does not drain the wetland he is providing benefits in the form of improved amenity attributes, enhanced wildlife habitat, and essential ecological processes. Which view is correct?

---

4   Danger to the general population comes at a higher nitrate concentration level than for infants and pregnant women, so N^* represents a concentration at which certain segments of the population would need to take averting actions.

Note that the differing perceptions of this issue fuel much of the incoherence - and the political struggle - in environmental policy focused on the agricultural sector. Farmers, often under siege from environmental interests, believe that their stewardship of rural resources constitutes provision of environmental benefits to the larger public. For these they often believe that gratitude should be forthcoming, if not financial rewards. And we know that agricultural policy in many OECD countries does indeed provide financial rewards for farmers to undertake land-use practices that are protective of environmental attributes. Such payments are *prima facie* evidence that - at least in political terms - their actions are regarded as providing these beneficial effects. Another way to think of it is that these environmental benefits would not be provided by farmers unless payments would be forthcoming. In essence the farmer has a contract with the state to provide a bundle of environmental benefits in exchange for financial payments. Whether or not the beneficial effects are real is an empirical question.

The other side of the argument would suggest, however, that farmers should not have to be paid to provide those things which, in the absence of agriculture, would be the natural order of affairs. They would point out that agriculture in the American midwest has already destroyed the vast majority of wetland habitat. In addition, soil erosion and chemical contamination are companions of modern agricultural practices in much of the world. These individuals would point out that industrial polluters are often required to pay for their environmental implications. Why should not farmers face the same financial disincentives?

Can we shed light on these two different perspectives? There are two American legal cases that may help us to understand the issues here. In *Just v. Marinette County*, the Wisconsin Supreme Court (56 Wis. 2d 7,201 N.W.2d 761, 1972) upheld a local zoning rule that prevented the draining of wetlands without prior approval from the county government. The Justs started to drain a wetland but were stopped by the local sheriff. When the case reached Wisconsin's Supreme Court, the county permit requirement was upheld and the Justs were denied the right to drain their land on the edge of Lake Noquebay. The Court insisted that the Justs had bought land that was wet, and after the permit was upheld, they still owned what they had bought. The permit did not take anything from them that they ever owned - though they may well have supposed that they could drain the wetland and acquire yet more "land." The Wisconsin court ruled that the prohibition on draining the wetland did not call for compensation to be paid to the Justs under the U.S. Constitution's "takings clause" (the fifth amendment).

We see here an argument that goes to the heart of the benefit/harm distinction. The Justs did not dispute the county's position that the wetland along the shores of Lake Noquebay provided important ecological services to the Lake and its downstream watercourse. The Justs did argue, however, that by being denied the opportunity to drain some of the wetland, they were being called upon to "provide" environmental services to the larger Lake Noquebay system. That is, by keeping part of their property in wetlands, they were providing environmental benefits to society at large. This is the difference between H_F and H_S in Figure 3.

The Wisconsin Supreme Court took quite the opposite approach. The Court reasoned that the Lake and its surrounding wetlands had been there longer than had the Justs, or the county government, or indeed any human habitation whatsoever. On that logic, the Court argued that the natural state of affairs was for the wetlands to serve as part of the larger ecosystem. In that role, the wetlands provided nutrient filtering (and a sink) for chemicals that would otherwise harm the Lake and its downstream extensions. The denial of a permit to drain the wetlands did not mean that now, suddenly, the Justs were providing something valuable - the continuation of which warranted paying them compensation. The Justs, said the Court, were not providing a public benefit at all. Rather, the

Court insisted that should the Justs drain the wetland they would be visiting harm on the entire Noquebay watercourse. Hence they could not drain the wetland, nor were they to be compensated for this denial.

There is a second important case for untangling the distinction between providing benefits and preventing harm. In *Penn Central Transportation Co. v. City of New York* (438 U.S. 104, 1978), the New York City Landmarks Commission prohibited the Penn Central Transportation Company from erecting a very tall building on top of Grand Central Terminal (which was already owned by Penn Central). The Landmarks Commission argued that such a structure would destroy the aesthetics of Grand Central Terminal.

The U.S. Supreme Court ultimately upheld the decision of the Commission and the benefit/harm distinction played an important role in the Court's finding. Plaintiff (Penn Central) argued that to be prevented from erecting the massive structure was a "takings" of its property rights to the airspace above the terminal it already owned. If it could not build, it demanded compensation for the lost income the large structure would have brought. Penn Central pleaded that it was being made to provide the public benefit of an unmarred Grand Central Terminal, as well as providing the benefits of open space and light in central Manhattan; other owners had been able to erect monumental structures, why was it now being singled out?

The Court found in favour of the defendants (the Landmarks Commission), and in doing so recognised something very important. The Penn Central Transportation Company argued that if it was to be forced to "provide" the benefits of landmark preservation and open space to others in central Manhattan, then it should be compensated for the denial of its "right" to build. It was, or so plaintiffs argued, entitled to be compensated for the financial loss that it must incur in order to "provide" public benefits to the larger society.

The U.S. Supreme Court found otherwise. As with *Just v. Marinette County*, the Court insisted that the *status quo ante* was one in which Grand Central Terminal and the open space above it were already being "provided." Indeed, the building plans of Penn Central Transportation Company would deprive central Manhattan of precisely those beneficial aspects and this the Court would not allow. Nor was Penn Central to be rewarded with compensation for its inability to destroy those attributes regarded as an essential part of the *status quo ante*.

In both the wetlands case, and the Grand Central Terminal case, we see that courts understand the fundamental distinction between providing a benefit to the general public, and the prevention of harm. Turning back to agriculture, such recognition holds important implications for those who argue that they should receive compensation for "providing" a range of amenity, habitat, and ecological benefits. Of course the legal findings from America are not pertinent throughout the OECD countries. But the logic from these cases offers some insight that may inform the political debate in other places.

We have, in essence, two *status quo antes*. One might be thought of as the political *status quo ante*, while the other is an environmental *status quo ante*. On the political side, agricultural interests will argue that they are providing public benefits. There is a plausible case for this argument when we think of the amenity implications of agriculture. After all, it is the manicured countryside that many non-rural residents now associate with rural areas. As strange as it may seem to environmentalists, many urban residents may indeed hold the well-maintained agricultural landscape in higher esteem than they would a "natural" forest. If so, then agriculture provides net amenity benefits as compared to a more natural landscape. But if fences and hedgerows are ripped out to permit ever-larger machinery, and if quaint barns and other buildings are removed, then the amenity dimensions of agriculture probably suffer.

The same may well hold for habitat implications of agriculture. If the agricultural landscape is geared toward some provision of wildlife habitat, then agriculture may be a net provider of habitat benefits. We know that a varied landscape is ideal for a range of wildlife, and under the proper circumstances agriculture provides precisely that varied habitat. But of course when farmers undertake to drain wetlands, or to clear forests to expand the area in crops, or to homogenise the landscape in other ways, then the habitat implications of agriculture are negative.

As for the ecological implications of agriculture, there is probably less of a case for the beneficial aspects of agriculture. Of course there are agricultural regimes in which an effort is made to use nature as an ally rather than as an enemy. Sustainable agriculture tends to see itself in this light. But for the most part, "modern" agriculture - with its heavy reliance on chemical inputs - could not lay claim to that status. Here, the ecological case against agriculture is probably compelling.

We see that the idea of a political *status quo ante* and an environmental *status quo ante* is at the heart of much of the struggle over the environmental implications of agriculture. In some instances the difference may be quite obvious. If pure groundwater is contaminated by heavy applications of agricultural chemicals then it stretches credulity to suggest that farmers should be compensated for their inability in the future to carry on such practices. But of course the political power of farmers may be such that they are able to acquire compensation in order to agree to cease these practices. In this case we would say that they have managed to use the political system to define a new *status quo ante*, any change from which must be compensated.

But things are not always so easy. We cannot logically insist that the environmental *status quo ante* is always properly regarded as that prevailing prior to any human action whatsoever. There has been, after all, human activity in western Europe for a very long time and so little of that environment is "natural" or "pristine." This suggests that with amenity and habitat aspects of the rural countryside, a more nuanced approach is necessary. Recall that in Figure 4 the conflict was between so-called "experts" and farmers about the acceptable level of nitrates in groundwater. However, there was certainly no dispute that nitrate concentrations prior to "modern" agriculture were very low indeed. The issue with nitrates is simply one of "acceptable" standards.

Notice, however, in Figures 2 and 3 that the issue is not one of acceptable levels — a threshold — but rather of more general standards of landscape appearance and habitat attributes. Of course there might be disputes about these things as shown in the Figures 2 and 3. But the concern is not to compare the current situation with some pre-human condition. Rather, the concern is to determine in the political arena what level of amenities and natural habitat is regarded as the acceptable reference level against which deviations are to be penalised or rewarded. We see an environmental *status quo ante* and a political *status quo ante* at work here, but in a more subtle way than with nitrate contamination.

But the ultimate test is found in what we regard as the environmental *status quo ante*. Whether or not the general conclusions intimated above really hold up to scrutiny will depend on whether or not the perceived "natural state of affairs" has been improved upon or diminished. So we see that there can be fundamental disagreements about whether particular actions constitute the provision of benefits to the public, whether the benefits would be forthcoming regardless of the efforts of farmers, whether agriculture represents a net decrease in the three environmental implications (amenities, habitat, and ecological processes), or whether the actions of farmers mean that important damages are precluded. And these disagreements lie at the heart of many environmental disputes in agriculture. The resolution of these disputes is found in the particular property rights regime in place at the moment. Or, the resolution depends on a change in the presumed property regime. To that I now turn.

3.   Property regimes in the benefits/damages debate

The previous discussion reminds us that much of the disagreement over the correct policy response to the environmental implications of agriculture arises from the different perceptions about property rights inherent in land. Farmers will argue that since they own the land on which they farm, they are free to treat it as they wish. If a wetland stands in the way of greater agricultural production, then the farmer believes that he has a "right" to pursue that production. Wetlands are an impediment to greater agricultural production just as forested areas were impediments in earlier times. In the farmer's mind, the wetland inhibits higher production and income and so is seen as a liability.

Notice that this matter is not confined only to wetlands. As suggested above, trees can also be seen as impediments to agricultural production. Indeed, agricultural history in Europe and North America has been a constant struggle against the forest. Agriculture keeps the forest at bay. A balance seems to have been struck between the proportion of the landscape that shall remain forested, and the proportion that shall come under the plough. But forests are not as scarce as wetlands in most areas and so a new controversy persists.

Those who care about rural amenities, rural habitat for wildlife, and general ecological processes will argue that the mere fact of land ownership does not bestow the right to destroy nature. They would point to a range of land uses that are no longer permitted, even though at one time - under different socio-economic conditions - those particular uses were thought to be acceptable. The advent of town and country planning in Britain following World War II, and indeed land-use controls throughout much of western Europe, reflect these changing social perceptions of "ownership."

In America, even though land-use controls have not reached European proportions, urban zoning has long ceased to be controversial. While we find in America a resurgence in the so-called "property rights" groups that challenge environmental regulations, these reactions are, in all likelihood, a sign of the declining acceptance of the view that ownership of land carries with it some automatic "right" to disregard the larger social interest in land and how it is used [Bromley, 1993; 1995][5]. If property rights were not undergoing transformation, there would be no need for such groups to adopt aggressive tactics.

So the question becomes one of addressing the role of property rights in making sense of the distinction between providing environmental benefits as opposed to preventing environmental harm. Different OECD countries have their own particular rights structures with respect to land and related natural resources. For current purposes it will be sufficient to discuss property rights in a general way.

---

[5]   The Republicans who came to the U.S. Congress in the 1994 elections brought with them a particularly vigorous attack on environmental regulations. They have since introduced legislation that would require governments to pay compensation when governmental actions or regulations diminished property values by some percentage (sometimes the diminution threshold was as low as 10 per cent). A number of state legislatures have seen similar efforts attempted there. So far, very few of these bills have actually been enacted into law.

The obvious starting point is to confront the most extreme position regarding property rights. The natural rights doctrine, derived from the Lockean labour theory of property, holds that individuals acquire rights in land by virtue of having mixed their labour with land [Becker, 1977]. Because humans own their labour power, this self-ownership necessarily precedes civil society. On this basis no state may transgress the "property" of its citizens. We might think of this doctrine as being central to the idea of individual freedom. In other words, if the individual is sacred on natural rights grounds, then governments violate this freedom at their peril. From this foundation it is easy to take the next step which holds that the land and possessions of a free individual - whose freedom stands prior to the state - are absolutely protected from encroachment by the state. In its romantic version, the individual landowner stands immune to the tyranny of governments.

This logic has come under criticism on two fronts. The first is that individuals are not quite the free agents we imagine them to be. If natural rights really existed there would be no need for governments to protect its citizens from each other (and from the state itself). So it is only the authority system we know as governments that can give empirical content to the idea of rights for individuals. Rights only have meaning when there is an authority system to impose duties on others. While the idea of natural rights - and human rights - make fine political rhetoric, rights only have content when they are accompanied by duties on those who would interfere with us. And those duties can only come from a state that agrees to compel civil behaviour on the part of its citizens[6].

The second flaw with the Lockean position on natural rights and freedom arises from the problem of imperfect (incomplete) acquisition. That is, if owning property is to stand as the guarantor of freedom and protection from an arbitrary state, then we find ourselves with a protection of freedom that is at once incomplete and fickle in its application. Whence does the guarantee of freedom for the propertyless arise? In other words, it is incoherent to insist that property rights are the *sine qua non* of liberty when many among us have no property. Does this mean that liberty falls only to the propertied classes?

In addition to these two flaws, Locke himself added a famous - but frequently overlooked - proviso in discussing his theory of property. He insisted that acquisition by labour was only justified if there "was enough and as good" for others. In a world of scarcity, where there is not enough to go around, the labour theory of acquisition fails us completely [Bromley, 1989].

The difficulty in agriculture/environment interactions is that property rights claims are never far from view. While the claims of landowners will often be in terms of so-called natural rights and freedom, property rights are justified by purpose, and limited by necessity [Christman, 1994]. To quote R.H. Tawney:

> "The state has no absolute rights; they are limited by its commission. The individual has no absolute rights; they are relative to the function which he performs in the community of which he is a member, because unless they are so limited, the consequences must be something in the nature of private war. All rights, in short, are conditional and derivative, because all power should be conditional and derivative. They are derived from the end or purpose of the society in which they exist. They are conditional on being used to contribute to the

---

[6]   The clearest evidence of this issue is found in slavery. If individuals really had natural rights then slavery would have been impossible. States first sanctioned slavery, and later outlawed it. It is the state that has the legal capacity to define one individual "slave" and another individual "owner". Therefore only the state can eliminate those categories.

attainment of that end, not to thwart it. And this means in practice that, if society is to be healthy, men must regard themselves not as the owners of rights, but as trustees for the discharge of functions and the instruments of a social purpose [Tawney, 1948]."

The idea of property rights as derivative and limited is well accepted in most OECD countries. But visions of property rights as some absolute construct still loom large in public policy discourse concerning the environment. It is this idea that underlies much agricultural policy where anti-social land-use activities are often necessarily modified by the payment of inducements to farmers to adopt alternative practices [Bromley and Hodge, 1990]. Such payments serve to reinforce the idea that farmers have a right to allow top soil to wash away, to drain wetlands, and to apply toxic chemicals. The myth of absolute property rights affirms certain environmental practices, which must then be "bought out." And of course the act of "buying them out" further reinforces the myth of absolute rights.

We see this at work in the very different treatment accorded to industry and agriculture in the realm of responsibility for environmental practices. Some might suggest that industry is treated less graciously than agriculture because of the economic power of the former, and the "perfectly competitive" nature of the latter. A more plausible explanation would concern the political sentimentality of all things agrarian. And of course the Lockean imperative is at work here as well. But there is also the fact that industrial pollution tends to come from the end of a pipe (or a smokestack), while agricultural pollution is more diffuse in its origins. Despite the diffuse nature of agricultural pollution, policy options do exist for addressing the problem. In particular, one solves non-point-source pollution problems by forming associations within particular watersheds and making the group of farmers collectively responsible for water quality. If pollution fees are levied, they are assessed against the collective as a group. This then forces the individual members of the group to monitor each other's behaviour, and to assess miscreants accordingly [Bystrom and Bromley, 1996].

The practical effect is that agricultural practices that harm the environment are seen as part of the farmer's rights bundle. On this view, such practices can only be stopped with proper financial inducement. In other words, when farmers agree not to drain wetlands, or agree to reduce chemical applications, or agree to contouring practices that will impede the runoff of precious soil, such actions are interpreted as an example of civic-minded behaviours that will suddenly provide valuable environmental benefits to the community at large. For this, the case is then made that cost-sharing is appropriate. Or perhaps some other form of financial beneficence ought to be forthcoming [Bromley and Hodge, 1990]. Notice that if property rights were seen as less absolute than the Lockean myth suggests, then the harm that such practices create could be halted with impunity - and at much lower cost to the public purse.

By way of summary, property rights bestow on the right holder the capacity to compel the state to stand behind the interests of the party with rights. Hence, the state - through its interest in all transactions - chooses to protect certain parties and their interests as against the interests of all others. This means that coherent rights can only be established through a process that starts not with physical acts, but with Kantian reason [Bromley, 1991]. And this means that all rights are constructed because they are wilfully determined through collective action grounded in Pure Reason.

4.   Missing markets in agricultural externalities

The quest for improved environmental policy is influenced by the extent to which the environmental implications of agriculture fall outside of the decision calculus of farmers. We have, in essence, a problem in which the physical implications of agriculture transcend the boundary of the firm responsible for those implications. When English farmers rip out hedgerows to allow for larger machinery, urban residents are outraged. When American and German farmers use nitrate fertilizers, drinking water is contaminated far beyond the boundary of the farm. When Asian farmers burn off rice stubble, non-farm populations are adversely affected.

We are reminded again that actions taken within the decision confines of the farm firm hold implications for those beyond the decision calculus of that firm. If beneficial environmental implications are forthcoming, say the provision of a more pleasing landscape than otherwise, or the provision of enhanced wildlife habitat, then a case might exist that payments should be forthcoming to farmers to pay for these off-site benefits. Indeed, agricultural policy in many countries probably does that already. Whether or not the "optimal" level of rural amenities and habitat provision is forthcoming is interesting, but such a determination is too difficult, both conceptually and empirically, to undertake with any confidence.

To the extent that modern agriculture is a contributor to degraded ecological processes then the flow of payments to agriculture must be modified accordingly. In a sense we might imagine a regime in which credits and debits are accumulated for the various environmental implications of agricultural practices in particular areas. A credit or two for amenity benefits, a debit or two for chemical contamination, and perhaps neither credit nor debit for habitat attributes. Each case would need to be assessed individually[7]. Regardless of how this accounting might work out, it must be seen as a much more elaborated scheme than the simplistic idea that the polluter should pay. For as we shall see, it is not always possible to ascertain which party is the "polluter."

5.   The polluter-pays principle

The heart of much environmental policy is the idea that the polluter should pay for the damages being caused. This is only logical. Unfortunately, a problem arises because the very idea of "pollution" is often unclear. So the clarity of a simple rule is clouded by operational realities. The alliterative charm - as well as the superficial equity - of the "polluter-pays principle" explains part of its appeal to policy makers. There is a sense of certitude to the idea that those who "cause" pollution should pay for it. The difficulty, however, is to ascertain precisely who "causes" pollution.

When residential developments spread into areas that are clearly agricultural, the assumed clarity of the polluter-pays principle disappears. After all, agricultural activity implies noise and dust. When a particular region is exclusively agricultural and the dust and odours spread through the vicinity, the idea of "agricultural pollution" is probably an oxymoron. After all, farmers expect to endure noise and dust as part of their livelihood. So the polluter-pays principle does not work here.

Yet when non-agricultural activities encroach on the rural landscape, the normal attributes of agriculture - noise, dust, animal odours - become "nuisances" (and thus "pollution") to the new inhabitants of the countryside. Does the polluter-pays principle now work? Not necessarily.

---

[7]   There is no assurance in this system that the wrong incentives would not result in economic waste and a rural environment that is without redeeming virtue.

The polluter-pays principle is the logical outgrowth of the Pigovian perspective on externalities. Pigou saw externalities - specifically pollution - as an activity by one party that harmed another. In its simple Pigovian manifestation, factories emit smoke and thus laundries have to wash their linens again. Coase countered this vision by suggesting that perhaps the laundry might be able, more cheaply, to find another way to dry its linens. Or, perhaps the laundry should move so as to avoid the smoke. On this view, the laundry "causes" the problem by locating too close to the smoky factory[8].

Coherence requires that two distinct issues be separated. First, there is the physical emission of smoke. Second, this smoke is transformed into "pollution" when a victim is within the realm of the emission. Only with the juxtaposition of two parties does the pollution become an "externality" of policy significance. If we momentarily overlook air quality for breathing, there is no pollution (no externality) until a laundry enters the picture. And, as above, when the region is entirely agricultural, noise and dust and odour - though present - do not constitute "pollution" and therefore an externality cannot exist (by definition). As we have noted elsewhere:

> "It is therefore impossible to advance a rule that can generally define even the quite narrow question of physical causality. What then is the basis for the Pigovian perception and its implied taxing rule? Is there a universal right to undisturbed natural circumstances? If the answer is a tentative "yes," is this right based on the idea that the cheapest solution to externalities necessarily follows from the way responsibility is construed? Or is the rule hampered by the idea that whoever was there first has the right? [Vatn and Bromley, 1996]."

The Pigovian approach and its taxing policy (the polluter-pays rule) seems intuitive because the most common pollution cases are ones in which the emitter was not there first but came after long-established uses were well underway. There have always been laundries - or families - hanging out linens, but smoky factories are the product of the Industrial Revolution. But by altering the sequence of things, the apparent clarity disappears. And this is the problem alluded to above when residential developments encroach on long-standing agricultural areas. Consider the famous English case of *Bryant v. Lefever*. Commenting on this case, Coase writes:

> "The plaintiff and the defendants were occupiers of adjoining houses which were of about the same height. Before 1876 the plaintiff was able to light a fire in any room of his house without the chimneys smoking: the two houses had remained in the same condition some thirty or forty years. In 1876 the defendants took down their house, and began to rebuild it. They carried up the wall by the side of the plaintiff's chimneys much beyond its original height, and stacked timber on the roof of their house, and thereby caused the plaintiff's chimneys to smoke whenever he lighted fires [Coase, 1960]."

So here the plaintiff, minding his own business, suddenly found that when he tried to light a fire he was forced from his home by his own smoke. But in this case the harm was "caused" by his neighbour who had raised the height of his house, thus destroying the draw of plaintiff's chimney. After all, in the *status quo ante* - before Lefever added to the height of his house - Bryant's chimneys worked quite fine. So Lefever "caused" the pollution newly experienced by Bryant. The appeal of this particular conflict is that it forces us to reassess our often-simplistic notions of "cause." On the standard Pigovian logic, the plaintiff had a convincing argument. It was not the smoke *per se* that

---

[8] Such logic often causes some to insist that the right should go to the party who was there first. But of course this time-dependent approach cannot always be relied upon. More will be said on this below.

was the problem, for Bryant had been lighting fires for a very long time. Instead, the harm was caused by the neighbour's new higher building.

As the case worked its way through the courts however, things became somewhat less clear. In the trial court, the plaintiff was awarded compensation for the defendant's actions. On appeal, however, the judgement was reversed. To quote from the appellate decision:

> "They (the defendants) have done nothing in causing the nuisance. Their house and timber are harmless enough. It is the plaintiff who causes the nuisance by lighting a coal fire in a place the chimney of which is placed so near the defendants' wall, that the smoke does not escape, but comes into the house. Let the plaintiff cease to light his fire, let him move his chimney, let him carry it higher, and there would be no nuisance. Who then causes it? It would be very clear that the plaintiff did if he had built his house or chimney after the defendants had put up the timber on theirs, and it is really the same though he did so before the timber was there. But (what is in truth the same answer), if the defendants cause the nuisance, they have a right to do so. If the plaintiff has not the right to the passage of air, except subject to the defendants' right to build or put timber on their house, then his right is subject to their right and although nuisance follows from the exercise of their right, they are not liable [Coase, 1960]."

So the matter of "cause" turns out to be a tricky one indeed. The easy response is to denounce the appellate court as wrongheaded. But something quite profound is at work here. The trial court seemed convinced by the innocence of plaintiff building a fire in his own home. On the other hand, the appellate court seemed convinced by the need for "progress" and continual change in the *status quo ante*. If defendant was investing in his dwelling, making it larger and more habitable, how then can one quite inadequate chimney be allowed to stand in the way of modernisation? Perhaps the plaintiff [Bryant] had under-invested in chimney height when he built his house?

The trial court regarded the "harmed" party as the plaintiff, while the appellate court responded by saying that if the defendant must pay the plaintiff, then the "harm" really falls on the defendant. If the owner of the fireplace "causes" the externality by lighting his own fire, then it hardly qualifies as an externality in the eyes of conventional economic theory. The appellate court saw the essence of the problem in a way that the trial court missed. We see that the application of simple rules about "cause" and "harm" will fail as the grounds for a procedure to determine the appropriate rights structure.

Indeed, environmental policy is often confused because we search for just such a rule. Of course existing activities may give some presumptive claim to a "right," but this claim is just that - presumptive and tendentious rather than substantive and durable. Such presumptive rights are illusory until recognised in some institutionalised arena of affirmation and ratification. This arena could be a legislature, or a constitutional court. The chimney case (*Bryant v. Lefever*) reminds us that the actions of both parties are of singular importance in externality discussions. In essence, both parties were the "cause" of their new conflict.

An American case (*Spur Industries v. Del E. Webb Corporation*) illustrates another aspect of "pollution" with a direct bearing on agriculture. Spur Industries was engaged in the feeding of large numbers of cattle some considerable distance north-west of the fast-growing metropolitan area of Phoenix, Arizona. When first constructed, the Spur feedlots were very far out in the desert. However, Del E. Webb Corporation undertook the development of a housing complex in the immediate vicinity of the feedlots. The flies and odours associated with the feedlot were a serious impediment to the sale of Webb's homes. The Webb Corporation sued Spur Industries for creating a nuisance.

The court found in favour of the plaintiff (Webb) on the grounds of a genuine nuisance that did indeed harm Webb's financial prospects. But the court ordered Webb to pay the costs for the feedlots (Spur Industries) to move to a new location where the inevitable accompaniments of confined cattle feeding - flies and odour - would not constitute "pollution". Again we see that "pollution" is situation-specific.

We also see in *Spur Industries v. Del E. Webb Corporation* an attention to both efficiency and to equity. On the efficiency front, it is probably more important that residential developments be accommodated near other urban areas than that particular feedlots survive there. While the feedlots of Spur Industries might have been out in the desert when first constructed, it is not unreasonable to suppose that residential developments might invade the desert in such a fast-growing location. On the equity front, Spur Industries seems to have been victimised by the problem of Del Webb "coming to the nuisance." But the court made Spur whole by requiring that plaintiff (Webb) pay for Spur to relocate.

Of course the circumstances of this particular litigation would have been avoided in many countries where rural zoning is more developed. Spur would have been a permitted use where it was, and Del Webb would not have been allowed to approach the feedlots with housing. With zoning, the presence of flies and door would not have become a nuisance (pollution).

In essence, Coase was commenting on the fact that a strict Pigovian rule might impose costs on the "wrong" party. That is, sometimes the net social dividend would be enhanced if the "victim" were to bear responsibility for solving the conflict. In *Bryant v. Lefever* the "victim" was the plaintiff who merely wished to light his fire without suffering his own smoke. In the name of progress, the "victim" had only to expend a small amount to extend his chimney to restore its draw. This would surely be preferred, in the long run, to preventing Lefever from expanding his house and adding to the capital stock of the city. But of course, in the name of equity, Lefever might have contributed something to the costs of Bryant's taller chimney. This would be the solution taken from *Spur Industries v. Del E. Webb*.

It may be noticed that neither of these cases can give much comfort to agriculture. First, it is not always the case that "first in time means first in right." The history of the law as instrumental social policy tells us that those things defined as "progress" will inevitably win out over those things which are seen as preserving the status quo. The law is, after all, part of the evolving institutional structure of a society. For the most part, nation-states see their manifest destiny in terms of economic progress. And we must recognise that economic progress implies change.

Second, instances in which other uses (residential developments) "come to the nuisance" (flies and odour) are not always resolved to the satisfaction of the party who did not move. It is possible, through careful zoning, to separate mutually incompatible activities. But over time, it is probably true that activities that emit what will come to be regarded as "pollution" are living on borrowed time in that particular location.

Third, the idea of "cause" can sometimes be counterintuitive. In the stylised Coasean world of bargained solutions to externalities, the only relevant cost in considering judgements about responsibility for action is the level and incidence of transaction costs. Liability for remedial action must lie with the party best able to handle change with the minimum of such costs. Transaction costs are either so high that no change in outcomes is deemed "efficient," or transaction costs are low enough - at least for one side - to permit a bargained transaction. In this case, the cheapest solution will be found through bargaining. But the real world is rarely the idealised Coasean world. After all, in the real world, citizens, legislators, and judges insist that it matters who was there first, and it matters who has the higher moral claim in a conflict. That is why a simple rule about who is "causing" an environmental problem will usually fail us.

6.   Towards a policy framework

The policy context of agriculture

The subject of interest here is the "environmental implications of agriculture." Unfortunately, posing the question this way overlooks the fact that "agriculture" cannot be defined without reference to the policy context in which food and fibre are produced. After all, "agriculture" *per se* cannot possibly benefit or harm the environment because the term is devoid of the necessary specificity. Rather, agriculture in a particular place, using particular technology, engaged in particular enterprises, and shaped by particular economic incentives operating on farmers (often called agricultural "policies") gives rise to a particular constellation of economic and environmental implications - some of which will be regarded as beneficial by a subset of the population, and some of which will be regarded as harmful by a different subset of the population. Therefore, we simply cannot talk of the "beneficial effects of agriculture on the environment." All we can discuss is the environmental implications of a specific kind of agriculture.

The essential component in any assessment of the environmental implications of agriculture is the political and economic milieu that gives "agriculture" its empirical content. We must, therefore, focus our attention on the "policy context" of agriculture, not merely on the agricultural sector in the abstract.

All economic activity - whether industrial or agricultural - operates in a constructed economic context by which I mean the constellation of prices, laws, costs, technologies, and environmental circumstances that combine to produce a particular agricultural product. As we know, the process of creating some agricultural product also gives rise to other processes and phenomena. The mere existence of agriculture results in flows of income into and out of rural areas that, in many countries, is the social and economic foundation of the non-urban economy. But this contribution of "agriculture" to a particular sub-national region must be understood as an artefact of the political and economic milieu within which this production of food and fibre is embedded. To make the point, imagine how "agriculture" in western Europe would be structured - and how it would function (and look) - with a long history of American agricultural policy rather than with the CAP and the various national refinements thereto. And, as we see from the policy reforms over the past decade in New Zealand, "agriculture" in 1996 is certainly very different from what it was in the early 1980s.

The nature of agricultural policies

The policy climate for agriculture must be understood to entail three levels of interaction from the political process. At the most benign level, policy operates to facilitate certain behaviours by farmers that seem to be in the collective interest. I call these "facilitative" policies because farmers themselves also find them attractive but somehow unattainable. The most obvious of these policies are those that support commodity prices (or protect domestic agricultural markets) and hence generate farm incomes above levels that they would otherwise be in the absence of government programs. The policies must, by definition, satisfy some collective political goal (or else they would not exist), and these policies certainly are in the interest of farmers who benefit from them.

Turning to the environmental domain, another example would be better information about weather and soil moisture conditions to permit more precise applications of nitrogen fertilizers. Such precision would allow the farmer to reduce expenditures for fertilizers, and would also reduce nitrate leaching into groundwater. In the absence of this information, farmers have an incentive to over-use fertilizers on the principle of "better safe than sorry". Agricultural policies geared in this direction would save farmers money, and would probably reduce nitrogen contamination of groundwater.

At the next level of severity, policies can induce farmers to behave in new ways. Here, unlike with "facilitative" policies, we must deal with the fact that farmers see no particular benefit from undertaking these new behaviours. A tax on a particular chemical would reduce its use by farmers and this may leave their crops exposed to certain pests. Policies that induce new behaviours represent a degree of unwanted intrusion into the decision space of the farmer.

At the most extreme level, some policies will compel farmers to behave in new ways. A ban on a particular chemical, or a law prohibiting a particular production practice, or a law prohibiting the draining of a wetland introduce compulsion to the farmer's choice domain. Policies that compel are the least favoured approach from the farmer's perspective, as well as from the perspective of the state. After all, compulsion tends to alienate the compelled, and therefore may entail very high enforcement costs.

Value for money

The enduring mystery of agriculture concerns the profound political influence of farmers in the industrialised OECD countries where farmers are a distinct minority. A long list of explanations (rationalisations) exist for this phenomenon and now is not the time to review that literature. The practical effect is that agriculture has been somewhat indulged and favoured in comparison with industrial activities. There may be good and sufficient reasons for that treatment in certain countries. The important point here is that we must be careful to discern whether or not existing policies give good environmental value for the level of expenditures. I have written elsewhere that the presumptive property rights that attend land and agrarian pursuits have combined to distort economic incentives in perverse ways [Bromley and Hodge, 1990]. Farmers are often paid not to destroy nature; or they are given financial inducements to do those environmental things that other sectors would be compelled to do.

But as should be abundantly clear from my earlier comments, environmental goals and objectives (and instruments) remain the province of individual nation-states. There are, of course, certain trans-national imperatives, and so we find some general movement toward uniformity. And the new trading regimes under the World Trade Organization will impose yet another layer of uniformity on agricultural policies.

Within these international regimes, individual nation-states are obviously free to mandate whatever policies they wish. Draining of wetlands can clearly be prohibited with no compensation forthcoming. Agricultural chemicals can obviously be limited or restricted as a country wishes. Farmers need not be compensated or rewarded for good environmental practices. But that is a decision for each country to work out on its own. My remarks here are intended to remind the reader that farmers will, perhaps, hide behind the shield of "property rights" to justify their current practices - or their desire for compensation. Of all the OECD countries, the United States probably has the most stringent protections of "property rights." Yet even here, agricultural policies have managed to: (i) prohibit a range of agricultural chemicals; (ii) develop aggressive policies to protect soil, wetlands, and groundwater; and (iii) link price supports to conservation behaviour of farmers. It surely is easier to do these things in other OECD countries if the political will is present.

The dynamic dimension

We must always remember that the future is more important than the present. That is, agricultural policies must be evaluated for the incentive effects more than for the immediate response from farmers. By the incentive effects I wish to call attention to the fact that policies (agricultural or otherwise) redefine the choice set of economic agents. If fertilizers are more expensive, two things happen: (i) farmers use less; and (ii) there begins a search for a substitute. All prices - indeed all legal-economic circumstances - contain a static effect and a dynamic effect. Policy makers, and policy analysts, too often ask about the immediate response to particular policy reforms when the long-run (dynamic) effects are the more profound.

Another dimension of the long-run incentive effect has already been discussed. Farmers will argue that they have a "right" to do something and will insist that a denial of that "right" calls for compensation. The danger here, of course, is that once compensation is forthcoming, the tendentious "rights claim" of farmers has been politically reinforced. Again, we see that the long-run is more important than the immediate situation. Governments should avoid those policy actions that may bestow future rights (entitlements).

Whither optimality?

I observed above that determining the "optimal" level of rural amenities, habitat, and ecological processes was a most difficult conceptual and empirical task. But it is clear that economic policies in general — and agricultural policies in particular — hold important implications for the environmental aspects of the rural countryside. The policy challenge is to make sure that governments are not paying for behaviours on the part of farmers that should properly be considered as normal good stewardship of environmental resources. At the same time, it is necessary to understand that agriculture is no longer simply an activity that produces commodities for local, regional, national, or international markets. That has been the historical role of agriculture, but it is not the contemporary (or the future) role of agriculture in the developed world. Indeed, in the OECD countries, commodity abundance, not commodity scarcity, is the norm.

Given commodity abundance, it is necessary to begin to see agriculture as primarily a land management activity that provides (and supports) rural livelihoods, and that happens also to produce some marketable commodities. But some essential outputs of agriculture are not, nor could they be, marketed as are the commodities of agriculture. This fundamental redefinition of agriculture allows us to escape the conceptual trap that seems to prevail in many discussions about the environmental attributes of agriculture. That conventional view holds that there is some normal structure of agriculture in each ecological setting which gives rise to some "natural" level of costs of production. This thinking then allows a seamless transition into a discussion of subsidies and "distortions" that contravene some inherent comparative advantage. Recent preoccupation with revising world trade arrangements has tended to reinforce such thinking.

Unfortunately for those who believe they can rather effortlessly spot such "distortions," the very idea of a distortion or a "bias" only has meaning within some prior definition of what is assumed to be "natural" or "inevitable." Too often, there will be a belief that some natural and unfettered market will reveal this idealised state of affairs. From this supposition, there will then begin serious discussions about comparative advantage in trade. However, the idea of comparative advantage is simply an artefact of a large number of natural and social constructs. Sometimes, the social construction of "comparative advantage" will be rather obvious. If the subject of discussion is automobiles then the purpose of the sector under study — autos — is rather straightforward. However, if the subject of discussion is agriculture, then the purpose of the sector under study is no longer so straightforward. That is, agriculture is no longer just about producing tradable commodities. Agriculture is now a multi-product sector in which simplistic ideas of efficiency, or of "subsidies," will mislead. In a world of commodity abundance and environmental scarcity, the old logic is both incomplete and inadequate. Simply put, the seeming clarity of "distortions" is — *ipso facto* — a phantasm.

The agro-environmental ledger

I suggested an accounting structure in which environmental debits and credits might be reckoned with respect to agriculture. There is not time here to develop this idea in great detail, but it seems a good idea and something like it is, perhaps, already underway in the European Community. There is now much interest in reconfiguring national income and product accounts to reflect environmental degradation. We might well imagine a similar, though elaborated idea, for the environmental implications of agriculture.

7.   Bibliography

BECKER, Lawrence C. (1977), *Property Rights: Philosophic Foundations London*: Routledge and Kegan Paul.

BROMLEY, Daniel W. and Ian Hodge (1990), "Private Property Rights and Presumptive Policy Entitlements: Reconsidering the Premises of Rural Policy," *European Review of Agricultural Economics* 17:197-214.

BROMLEY, Daniel W. (1989), Economic Interests and Institutions: The Conceptual Foundations of Public Policy Oxford: Blackwell.

BROMLEY, Daniel W. (1991), Environment and Economy: Property Rights and Public Policy Oxford: Blackwell.

BROMLEY, Daniel W. (1993), "Regulatory Takings: Coherent Concept of Logical Contradiction?" *Vermont Law Review* 17(3):647-82.

BROMLEY, Daniel W. (1995), "Rousseau's Revenge: The Demise of the Freehold Estate," Keynote address at the conference "Who Owns America?" Land Tenure Center, University of Wisconsin-Madison, June 21-24, 1995.

BYSTROM, Olof and Daniel W. Bromley, Contracting for Non-Point-Source Pollution Abatement, University of Wisconsin-Madison, Department of Agricultural and Applied Economics, Staff Paper No. 392, March 1996.

COASE, R.H. (1960), "The Problem of Social Cost," *Journal of Law and Economics* 3:1-44.

CHRISTMAN, John (1994), The Myth of Property Oxford: Oxford University Press.

TAWNEY, R.H. (1948), The Acquisitive Society New York: Harcourt, Brace and World, Inc.

VATN, Arild and Daniel W. Bromley (1996), "Externalities: A Market Model Failure," Environmental and Resource Economics (forthcoming).

ENVIRONMENTAL BENEFITS OF AGRICULTURE: EUROPEAN OECD COUNTRIES

by
Peter Lewis Nowicki
ECNC[9], Tilburg, The Netherlands

Introduction

Two concepts are an important framework for an overview of the environmental benefits from agriculture in Europe:

- The semi-natural habitats of Europe, which embody many of the biodiversity and landscape (amenity) values within rural areas, have been created as a by-product of a rural economy which has changed radically, and in which agricultural land use practices associated with many semi-natural habitats have themselves often fallen out of use.

- The cessation of agricultural activity - either extensive or intensive - in any particular locality will not necessarily result in the restoration of biodiversity and landscape (amenity) values that have been present at some former time in history; other influences, linked to ecological dynamics and agricultural land use practices themselves, are guiding natural successionary processes, and these regulators change in nature and in degree over time.

For millennia, man has used the natural resources at his disposal, by taking advantage of numerous opportunities to benefit from nature. The branches of pollarded willows lining dikes and water courses have been used for tool handles, their roots providing stability for the embankments. The creation of terraced fields, ringing hillsides, by the artificial levelling of the soil maximises water retention, creating arable surfaces when none previously existed and at the same time using the rocky material which would otherwise hinder cultivation for structural support. Hundreds of landscape attributes are an outcome of the parsimonious frugality by which humans have been able to improve their productive capacity. This use of natural resources even contributed to the evolution of plants and animals as species moved in to colonise new ecological niches and were exposed to subsequent evolutionary pressures at the same time as human-modified landscapes took their attributes, in a process referred to as co-evolution.

---

[9] The European Centre for Nature Conservation (ECNC) is a network organisation, linking together research institutes and conservation agencies in order to address the integration of the conservation of nature and especially of biodiversity within the European economic and political context. It is in this light that the discussion on the "overview on the environmental benefits from agriculture in Europe" is written, as it combines information coming from 11 scientific units in 8 eastern and western European countries with information taken from the literature.

As a result of human activities, the original, relatively homogeneous, forest cover of Europe became interlaced with tilled fields and pasture, hedgerows and orchards, to create a few thousand specific biotopes and hundreds of large ecosystems. The interplay of several hundred soil types (many resulting from human-influenced pedogenesis), climate, available gene-pools, and human use of the land has determined the current distribution in Europe of more than 250 species of mammals, some 520 species of birds, a 200 species of reptiles, some 70 species of amphibians, about 230 species of fresh water fish, 12 500 vascular plant species and 200 000 invertebrate species. Many of these species are confined to particular ecological regions, such as the boreal forest, the open field agricultural system, alpine ranges and riverine or estuarine wetlands [Nowicki, 1995].

1.   The development of European agrarian systems

Agriculture, as understood today in its double form of cropping and livestock raising, had its origins in the Neolithic period, around 8500 BC, in the Near East. As agriculture spread north and to the north-east, it displaced the hunting-gathering system in radical transitions in Greece (6000 BC), Italy (5000 BC) and central and eastern Europe up to the Rhine (4000 BC); a slower co-optation of agricultural methods by hunter-gatherer societies took place in France and Spain (4000 BC) and within the British Isles (3000 BC). This progressive spread of agriculture favoured adaptations in farming systems that corresponded to regional environmental differences, both in climate, soil structure and topography. Six broad types of agrarian systems have evolved in Europe, of which five are still present in modified form today.

The first of these agrarian systems occurred in proximity to the origin of the Neolithic farming technology. The eastern Mediterranean region took over the three-component Neolithic agrarian system without major change, with the first component being a three-stage cropping pattern of winter cereals (wheat and barley) followed by millet and then a long fallow period, making a two year cycle. The second component is a permanent crop, either orchard (primarily olive, but also nut and fruit trees) or vineyard; the orchards are inter-tilled in the first years before the tree canopy spreads significantly. The third component is small livestock grazing in unproductive, especially hilly, areas. The western Mediterranean region has a variant on the livestock raising, which is long-distance transhumance between infertile lowland and highland areas in the winter and summer seasons, respectively. The lower lowland fertility meant that an arable crop could be obtained only once every five years on the average, leaving larger areas available for extensive fallow pasturing. The principle of inter-tillage becomes more significant as a predominant landscape feature in the dehesa (open oak woods).

The second farming system, associated with the plains north of the Mediterranean, is the three-field system, the major variation being that wheat is the principle cereal in the warmer areas of England, France and north-western part of the Iberian peninsula, and rye - better adapted to colder climatic conditions - being the principle cereal in Germany, Scandinavia and Poland. Oats and barley are found throughout the three-field system, principally either for livestock feed or as a base for beer. The distinction with the Mediterranean agrarian system is two-fold: the sequence of winter and summer cereals is reversed, allowing a longer fallow period (three year rather than two year rotation), and larger livestock, swine and in particular cattle, have more importance, requiring a hay meadow land reserve as a compliment to grazing pasture. The permanent orchard crops associated with this system are apple, pear, peach, cherry and plum; regional aptitude by certain fruit (e.g. apple in Normandy), gives rise to local products associated with the area (in this case, cider and calvados).

The third farmer-herder system overlies the Mediterranean and three-field agrarian systems, in which shepherding and transhumance is associated with a simple in-field/out-field cropping pattern, in which the in-field is largely devoted to livestock feed cereals and garden vegetables and the outfield used basically for pasture. This system is found in the hills in contrast to the other two which are associated with large river valleys and plains.

A fourth system, no longer practised since the beginning of the twentieth century, but which has had an impact in the forest composition of Scandinavian countries, is a type of clear-cutting regime associated with the sowing of barley and rye. The practice was a two-stage cycle that existed principally in Finland and for a short while in Sweden: the first, required three years of preparation for a fertile crop-base in the fourth year, after which the parcel was abandoned to forest regeneration for twenty to thirty years before burning and planting within a single year.

A fifth system, nomadic herding, comes from Asia, and is typical of the south-eastern steppes and Hungarian Puszta, as well as the sub-arctic areas in northern Scandinavia. The sixth agrarian system is related to citrus fruits, arriving in Europe two thousand years ago and requiring an intensive use of human labour as well as of irrigation for maintaining orange and lemon groves. This system has less of an impact over a wide area than the other systems, and is associated with the Mediterranean region.

European agriculture has evolved on the basis of these agrarian systems. New crops, principally maize and the potato, but also squashes, pumpkins and beans, have been imported from the Americas, and were integrated in a geographic sense according to favourable climatic conditions followed by an expansion due to genetic selection. In general, the geographic distinction in agrarian systems has become blurred or has disappeared, either because of field engineering or genetic manipulation. Vestiges of former systems remain in the more infertile or intractable areas for field engineering (largely associated with hills and mountains, sometimes with soil quality as in the case of heath and moorland). Specialised produce, as for example the wide variety of cheeses and wines associated with particular grazing regimes or soil attributes, respectively, remain an important feature of European agriculture. A second important development in historical terms is market gardening around the increasingly populated urban areas.

The increasing opportunity to use machinery and genetically modified plant material in agricultural practice has given rise to a geographically significant spread of cash crops (principally wheat and barley, but also rice in estuarine locations to the south) accompanied by large-scale changes in field structure. The development of field engineering technology, since the time that windmills were first used for water evacuation at the beginning of the fifteenth century in Holland, has allowed large areas of Europe to be drained as well as irrigated. Another Dutch innovation, which has been enhanced through the increasing capacity for water evacuation by wind, then steam and finally electrical power, is the reclamation of land through polders; this practice spread rapidly along European coastal areas on the Atlantic seaboard in the sixteenth and seventeenth centuries. Thus the possibilities for producing particular agricultural commodities are no longer constrained to the historical zones of agrarian systems: it is revealing to compare the general outlay of the historical agrarian systems with that of recent potential for specialisation in particular agricultural commodities.

Table 1. Evolution of land use in Europe by country, 1960 - 1990, as percentage of total country area (land and water)

Country \ Year	Arable and permanent crops				Permanent meadow and pasture				Wooded area				Other land			
	1960	1970	1980	1990	1960	1970	1980	1990	1960	1970	1980	1990	1960	1970	1980	1990
Austria	20.9	20.0	19.5	17.9	27.4	26.4	24.3	23.8	37.5	38.2	39.1	38.5	12.9	13.9	15.7	18.4
Belgium	29.9	27.1	24.9	25.7	25.2	24.3	21.8	19.0	19.7	20.1	20.2	20.2	24.3	27.6	32.2	34.3
CSFR (former)	42.44	41.7	40.4	39.8	14.5	13.8	13.2	12.8	34.4	34.8	35.8	36.1	6.9	7.9	8.8	9.2
Cyprus	17.2	17.2	17.2	17.0	0.5	0.5	0.5	0.5	13.3	13.3	13.3	13.3	68.9	68.9	68.9	69.1
Denmark	65.4	62.1	61.6	59.7	8.0	6.9	5.8	5.0	10.2	11.0	11.4	11.4	14.8	18.3	19.5	22.2
Finland	7.9	7.9	7.6	7.2	0.3	0.4	0.5	0.4	64.4	66.2	69.0	68.7	17.5	15.6	13.1	13.8
France	38.8	34.6	34.3	34.8	23.8	24.3	23.3	20.6	21.1	25.4	26.5	26.9	16.1	15.4	15.7	17.5
Germany	35.7	34.7	35.1	34.8	18.6	18.6	16.8	15.7	28.6	28.6	28.8	29.1	15.3	16.2	17.4	18.2
Germany (east)	47.1	44.5	46.5	45.4	12.9	13.6	11.4	11.5	27.3	27.2	27.3	27.5	10.8	12.8	12.8	12.7
Germany (west)	30.7	30.5	30.2	30.1	21.1	20.8	19.1	17.6	29.2	29.2	29.4	29.8	17.2	17.7	19.5	20.7
Greece	28.0	29.6	29.7	29.6	39.5	39.7	39.8	39.8	18.7	19.8	19.8	19.8	11.4	8.5	8.3	8.4
Hungary	60.5	60.1	57.3	56.8	15.7	13.8	13.9	12.7	14.3	15.8	17.3	18.2	8.8	9.5	10.7	11.4
Iceland	0.1	0.1	0.1	0.1	22.1	22.1	22.1	22.1	1.0	1.2	1.2	1.2	74.2	74.0	74.0	74.0
Ireland	22.7	19.6	15.8	13.4	57.6	61.0	65.7	66.8	2.6	3.5	4.6	4.9	15.2	13.9	12.0	13.0
Italy	51.8	49.6	41.3	39.7	16.8	17.4	17.0	16.2	19.4	20.5	21.1	22.4	9.6	10.2	18.2	19.3
Luxembourg	29.2	25.4	22.7	22.2	24.9	26.7	27.6	26.6	33.4	32.1	31.7	34.3	12.1	15.4	17.6	16.4
Netherlands	28.4	23.7	22.0	24.4	35.6	36.2	32.1	29.4	7.4	8.1	7.8	8.0	21.6	24.2	29.1	29.1
Norway	2.6	2.5	2.5	2.7	0.5	0.4	0.4	0.3	20.4	24.4	25.7	25.7	71.2	67.4	66.1	66.0
Poland	51.7	49.0	47.8	47.1	13.3	13.5	12.9	13.0	24.8	27.3	27.8	28.0	7.7	7.6	8.8	9.3
Portugal	32.9	33.5	34.0	34.3	9.1	9.1	9.1	9.1	32.1	32.1	32.1	32.1	25.5	27.4	24.3	24.0
Spain	41.1	40.6	40.6	40.0	24.8	23.0	21.3	20.4	25.6	28.1	30.9	31.3	7.6	7.2	6.2	7.3
Sweden	7.9	6.8	6.6	6.3	1.5	1.6	1.6	1.2	61.4	61.8	62.0	62.3	20.6	21.4	21.2	21.6
Switzerland	10.2	9.3	10.0	10.0	42.2	43.3	39.0	39.0	23.8	23.8	25.5	25.5	20.5	20.3	21.9	21.9
Turkey	32.2	35.1	36.5	35.5	14.5	13.8	12.4	10.9	25.8	25.8	25.9	25.9	26.0	23.8	23.8	26.4
United Kingdom	29.8	29.4	28.6	27.2	51.1	47.6	46.9	45.7	7.0	7.7	8.6	9.8	10.8	11.8	14.7	16.0

Source: Europe's Environment Statistical Compendium (1995), Tables 8.3, 8.4, 8.5 and 8.6.

Figure 1. Causes of the decline of plant species in terms of the number of threatened plant species over 25 years

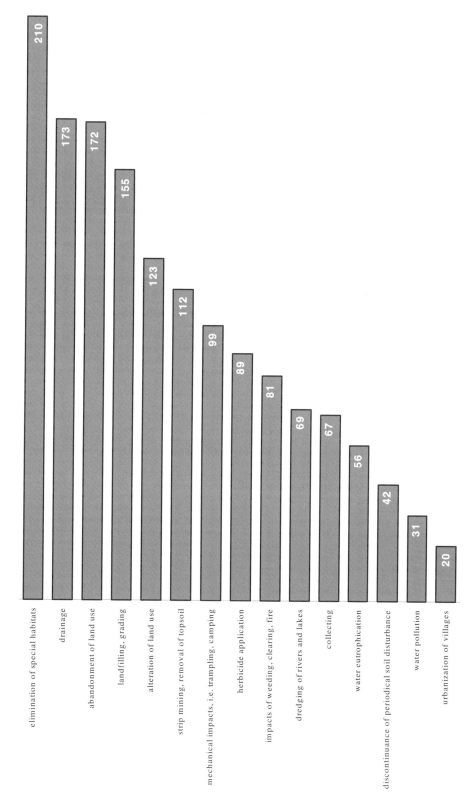

elimination of special habitats	210
drainage	173
abandonment of land use	172
landfilling, grading	155
alteration of land use	123
strip mining, removal of topsoil	112
mechanical impacts, i.e. trampling, camping	99
herbicide application	89
impacts of weeding, clearing, fire	81
dredging of rivers and lakes	69
collecting	67
water eutrophication	56
discontinuance of periodical soil disturbance	42
water pollution	31
urbanization of villages	20

Source: Council of Europe, 1987.

1. Note: Due to multiple listing of species, the sum of species in the diagram is greater than the total number of individual species in Europe (581).

2. Adaptation of agrarian systems to environment suitability and constraint factors

The geographical distribution of agrarian systems portrayed in the agricultural systems maps is only a partial representation of the influence of the present-day agricultural commodity market. Another approach is through an analysis of the land area for the different types of land use and the trends. Table 1 shows that in the last forty years a shift towards pasture or meadows for livestock and towards forestry is a general trend, but the table also demonstrates the relative importance of different land uses in terms of environmental suitability and constraints, and therefore also explains that biodiversity and landscape values as a consequence vary considerably throughout Europe. This variety in land use is also expressed in the geographical distribution of the larger landscape classes [Stanners and Bourdeau, 1995]. (Table 1).

Biological and technological innovation (including the invention of chemical fertilizers in Germany, followed by that of pesticides in the twentieth century) has allowed a greater quantity and variety of agricultural commodities to be produced to the point that the requirement for both labour and land has decreased, in an inverse relationship to the potential for capital investment to be profitable. This has led to the abandonment of certain agricultural practices that are labour intensive or which require access to large areas of land, and in some regions has led to the abandonment of agricultural land use altogether on land considered to be of marginal interest for more capital intensive agricultural land use [Bethe and Bolsius, 1995]. Human use of land can have direct negative effects on the biodiversity value of land, either through the elimination or fragmentation of natural habitats. But the abandonment of a particular land use can also have a negative influence. Figure 1 shows that the influence of various human activities as well as the cessation of activity through abandonment have a disruptive effect on vascular plants.

These changes in land use have had a significant effect on biodiversity and landscape values in Europe. Much of biodiversity has developed in association with the differentiated agricultural land use since the advent of Neolithic agrarian technology. This fact is represented schematically in Figure 2, and is also apparent from research into the biodiversity of specific farming regimes. What is apparent is that biodiversity increases according to the development of niches, or biological 'windows of opportunity', associated with agricultural land use. In Europe, the agricultural land use has a variety of regimes within the basic agrarian systems outlined above, which correspond to the great diversity of land forms and soil types within a relatively small area when compared to the American, Asian and Australian continents.

This relationship between agricultural land use and the diversity of land-forms and soil types has its expression in landscape features: these give rise to several types of cultivated and natural vegetation relationships referred to as cultural landscapes, for which one expert has denoted 30 classes covering Europe [Stanners and Bourdeau, *op. cit.*]; the CORINE Land Cover classification combined with the EU Habitats Directive nomenclature permits an identification of 36 rural land uses and vegetation types in a systematic way throughout the European Union [CEC, 1991]. What is certain is that the anthropogenic influence on the potential natural vegetation has resulted in a (non-urban) land cover that contains only agricultural or semi-natural land areas, with the exception of an extensive zone of virgin forest on the Polish-Russian border (Bialowieza), residual elements of virgin forest in Slovakia, and 'natural' vegetation in the mountainous or hardly accessible regions of Europe that has had little disturbance or has become restored through successionary processes after human land use has ceased.

What also is apparent is that the increasing importance of capital investment in agriculture has resulted in a negative impact on biodiversity and landscape values - a development that is related to greater homogeneity in agrarian commodity production, through an increasingly standardized production system, associated with an enlargement and conditioning of field structures that is appropriate for the use of specialised machinery and soil amendments [Jordan, 1996]. The paradox is that agriculture, in a historical perspective, has led to an increase in biodiversity in Europe before becoming directly responsible for its diminution [Tivy, 1990]. "In Norway, permanent old pastures managed without fertilizers may have 45 species of vascular plants per square metre, compared to 27 in abandoned old pastures that are no longer grazed, and 14 in fertilized, old pastures" [McNeely, in Halladay & Gilmour 1995, citing Götmark, 1992].

Figure 2. **Changes in plant diversity in Central Europe**

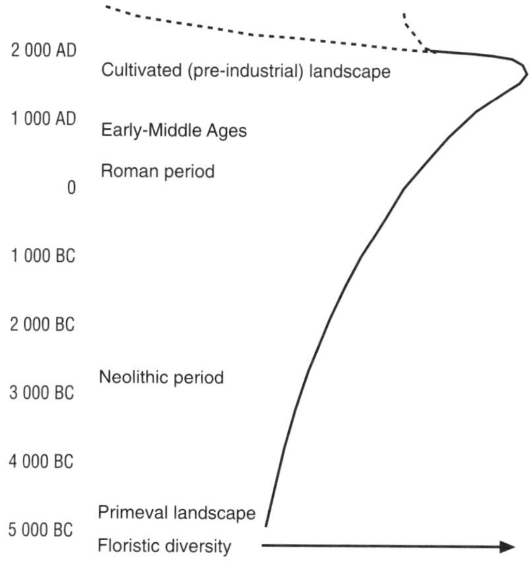

Source: Stanners and Bourdeau, 1995.

The expansion of arable agricultural land use has also been at the expense of some types of land cover. Forests have been cleared, either for pasture, meadows, and arable or permanent crops. Natural or human-created grasslands have also passed to arable land use. There have been several cycles of increasing and decreasing arable land use, linked to population fluctuations, agricultural trade considerations, and the market for timber (forestry is the primary alternative land use to agriculture) [Pounds, 1990]. The current phase is that of intensification of arable land use on less area, accompanied by an increase in the area devoted to a more extensive livestock management regime [van der Weijden, 1992]. It is possible that the intensification of arable land use is linked to the agricultural commodity market and that the extensification of livestock management is linked to financial incentives coming from the implementation of public-sector land use and market regulation policies [Tracy, 1993; Hoggart, 1995].

As seen in the preceding discussion, the development of agricultural land use has been as much under the influence of environmental factors as of market forces until the middle of the twentieth century, and the result has been a wide variety of landscape features which are the basis for an even greater biological diversity. The particular regional adaptation of agrarian systems can be understood as anthropogenic responses to suitability and constraint factors that even today remain valid parameters for sustainable agricultural land use.

3. Principles of sustainable agriculture

The notion of sustainable agriculture refers to an economically viable agricultural production system that maintains the natural resource base on which it depends, without negatively affecting environmental parameters (soil, air and water) or causing a socially undesired modification in landscape characteristics [after O'Riordan and Jäger, in Jäger *et al.*, 1995]; inherent in this notion is the idea of carrying capacity, which considers that the environment in which a human activity takes place can be more or less intrinsically suited for it.

The carrying capacity is not necessarily stable: short or long-term modifications of climate have occurred within recorded history (the Little Ice Age between 1200 and 1800), and may be occurring at present (a 2°C average temperature rise within the next 50 years). Nor are the markets - or subsidies - for particular agricultural commodities historically stable. Nor is the agricultural technology, including veterinary medicine, stable. Therefore the environmental aptitude and the profit margin of any agricultural pursuit will both inevitably change continuously, as will the by-products of agricultural land use such as biodiversity and landscape values. The component of change inherent in any agricultural system is underlined in recent research on sustainable agricultural systems, understanding 'sustainable' in the triple sense of the ability of such systems to remain productive in the long run from the biological/physical, economic and social perspectives [Barnett *et al.*, 1995].

4. Environmental benefits of agriculture

The six agrarian farming systems that were the traditional expression of agriculture in Europe were 'sustainable', in the sense that they lasted for several millennia; these systems, as discussed, became transformed in many geographical locations into the modern forms of agriculture, in which specialisation in particular commodity types has become a characteristic over large regions. These regional specialisations have accompanied a rationalisation of agricultural production in general, so essentially the basic commodity production of the traditional systems remains, but has become amplified in terms of output (as can be understood from a comparison of Maps 1 and 2).

The diminution of the biological and landscape values associated with the traditional six agrarian farming systems has led to research on their particular environmental benefits of some of their components which are still in practice today [Beaufoy, Baldock & Clark, 1994; McCraken & Bignal, 1995]. This research demonstrates the tight linkage between agricultural practice and the development of biodiversity values, and these are in turn often reflected in landscape values, not only at the field level, but throughout a larger area where such agricultural practices exist. An example is the difference in flora associated with differences in meadow management; Figure 3 schematically compares the diversity of species in the 'Nardetum' grassland found in the Haute Ardennes (Belgium) as management intensifies, and the legend also indicates the type of management along with the dominant floral species.

Figure 3. The relation between species diversity and the intensity of management with regards to grasslands

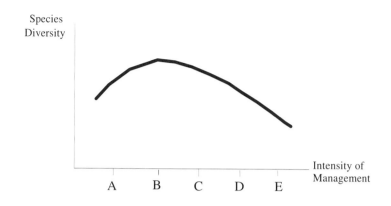

Source: after Beaufoy, Baldock & Clark, 1994, citing Peeters, 1993.

Legend:

A = extensive grazing
     Nardus stricta, Arnica montana, Pedicularis sylvatica

B = 1 cut/year for hay in rotation with extensive grazing, farmyard manure as fertilizer
     Meum athamanticum, Briza media, Potentilla erecta

C = 1 cut/year for hay and late season grazing, farmyard manure as fertilizer
     *Geranium sylvaticum, Sanguisorba officinalis, Polygonum bistorta,
     Alchemilla xanthochlora, Trisetum flavescens, Lathyrus montanus*

D = 2 cuts/year for hay, farmyard manure as fertilizer
     Anthriscus sylvestris, Heracleum sphondylium, Arrhenatherum elatius

E = 3 cuts/year for silage, livestock fecal slurry and NPK compounds as fertilizers
     Elymus repens, Poa trivialis, Taraxacum sp., Phleum pratense, Rumex obtusifolius

But such traditional practices are not necessarily 'sustainable' in the triple sense of the biological/physical, economic and social perspectives given by Barnett. It is for this reason that certain public-sector policies about land use and agriculture have particular provisions to encourage traditional forms of agriculture. Perhaps there should be some clarity about what the distinction between traditional and modern actually means.

Any development of categories by which to differentiate types of farming systems is a conceptually delicate exercise, is open to criticism. 'Traditional' agriculture practice can either be low-input (olive groves) or high-input (vineyards), either extensive (range grazing) or intensive (orange groves). 'Modern' systems of agricultural practice can be either 'conventional' (relying on chemical soil amendments) or 'organic' (relying on animal and vegetation based soil amendments). 'Modern' farming is increasing its profit margin through a careful regulation of inputs (quantity and timing of application), is experimenting with integrated pest-control, and is implementing a variety of soil conservation techniques, and also entails a reconsideration of what are appropriate types of mechanisation.

Thus there can not be a dichotomy between 'traditional' and 'modern' agriculture in terms of environmental benefits, the first having a positive connotation and the second a negative one; the question is that the array of environmental benefits will be different according to the land management practice, be this in terms of soil amendments and conditioning, cropping patterns (including fallow periods), stocking densities, and the way in which permanent crops are handled.

Map 1. **Historical agricultural systems**

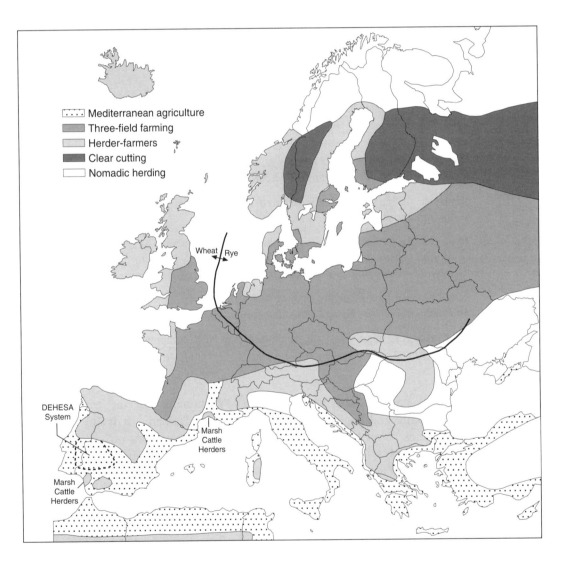

Ancient and traditional types of agriculture. The division between wheat and rye as the dominant bread grain is as of about A.D. 1900. Ancient and traditional systems gave way at widely different periods in the various parts of Europe. (Adapted from Jordan, 1996).

Map 2. **Present day agricultural specialisation**

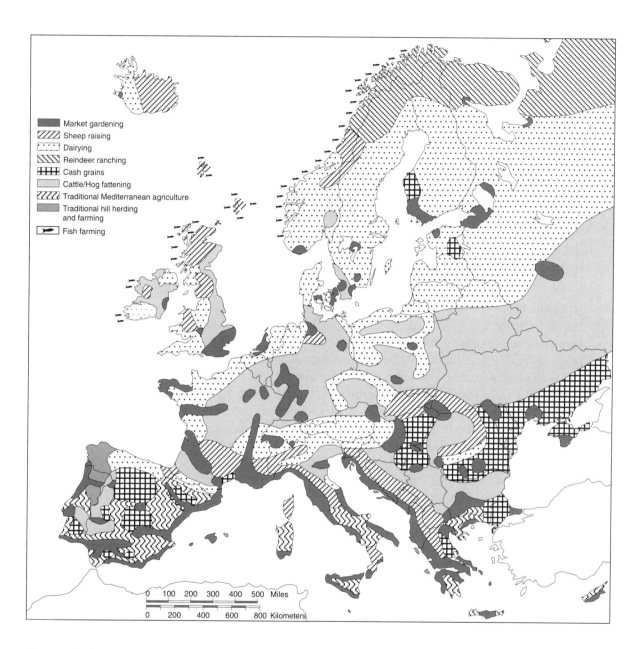

Source: Specialised types of agriculture in modern Europe. (Adapted from Jordan, 1996).

What is certain, however, is that particular forms of biodiversity and landscape values which depend on 'traditional' agrarian systems depend on the maintenance of very specific land management techniques, and these may no longer be associated with the evolution of the agricultural commodity market. That is because these techniques are adaptations to very specific environmental suitability and constraint factors. This can be seen in the case study in Box 1.

65

Box 1. Environmental benefits from agriculture in the Mediterranean region[10]

A series of closely related facts must be born in mind in order to understand the environmental value of agrarian landscapes in the Mediterranean and the problems they face at present. Basically, all of them are derived from a limited capacity of traditional technologies to alter the environment as well as from the (pre-)adaptation of Mediterranean landscapes to disturbance.

First of all, it must be born in mind that most Mediterranean landscapes we see at present are the result of human activities during the last millennia. The slowness of the anthropization process and the limited capacity of traditional technologies to produce great transformations of landscapes in short time spans have allowed for the preservation of a very valuable flora and fauna associated with them. This fact is also associated to the high disturbance regimes natural to Mediterranean ecosystems. In this way, two of the major human tools for landscape transformation, fire and herbivores, probably played a very important role in these ecosystems prior to human activities. The third human tool, its own labour and the work force of animals, probably had much less powerful effects, resulting in more local and slow changes.

Two examples can be used to clarify these ideas: dry-farming flatland (pseudo-steppes) and Mediterranean mountains. At present it is widely accepted that open lands occupied a large fraction of Central Iberian flatland and that fire and herbivores were probably responsible of it. Though fire initiation may be a rare event, the area affected by each individual fire over dry homogeneous woodlands should be very large. Therefore, the total amount of land affected by a recent fire and thus lacking a well-developed forest, was probably huge. The tendency of herbivores to graze on herbaceous communities together with the aridity of climate and low nutrient content of soils would further slow the development of woodlands in burned areas. The well-stated presence of steppe-birds in former times and the maintenance until present of plant species with a steppic origin also give evidence of the extent of open landscapes in the Iberian Peninsula prior to their anthropization. On this basis the impact of man can be understood as a progressive increment of landscape disturbance both in fire frequency and grazing intensity. Due to the non-existence of agro-chemicals, cropping would give rise to high-productivity landscape (*tessera*) from which wild fauna as well as ruderal plants would take advantage. Finally, the inability to plough the poorest soils gave a survival opportunity to other native species.

The situation in the Mediterranean mountains was probably very similar in relation to fire and herbivores, though the extent of burned areas would be smaller due to environmental heterogeneity, the presence of rock outcrops and the frequency of plant communities associated with the presence of water-courses (less fire-prone). Other human actions (terrace building, meadow improving, sheepfolds...) gave rise to habitat fragments with more developed and/or nutrient-rich soils, somewhat similar to those natural (though more scattered) in mountain areas. In summary, all these actions would increase the extent of high-productivity *tessera*.

(continued on next page)

---

[10] Contributed by Dr Juan E Malo and Dr Begoña Peco, Facultad de Ciencias, Departamento de Ecología, Universidad Autónoma de Madrid.

(continued)

Some characteristics of exploitation systems are also responsible for the high natural value of these agrarian landscapes. On the one hand, traditional exploitation was basically autarchic and thus based on complementary uses that took advantage of spatial and temporal variability. The orographic and edaphic complexity of Mediterranean areas reinforced these trends. On the other hand, and as a particularity of the Mediterranean region, intra- and inter-annual climatic variability limit the intensification process. And this limitation is determined by the stress seasons (summer in particular) and especially by the stress years (periodical droughts). Due to these facts, modernisation and intensification of agricultural practices have not had important impacts on many local agrarian ecosystems.

The major benefits of the existing agricultural practices are associated with the maintenance of a moderate and complementary exploitation through space and time. This is principally so for i) the maintenance of anthropic landscapes that can easily become destabilised if abandoned (i.e. terraces which avoid the erosion of soil accumulated for centuries), and ii) the continuity of low-intensity exploitation systems that allow for the conservation of species that rely on them (i.e. the avoidance of ploughing poor soils where many plants and animals are present) or even take advantage of them (i.e. insects and plants that grow in cultivated fields and are the main source of food for steppe birds).

Given the origin of the existing landscape, its continuity depends on maintaining an extensive agrarian system, which is not favoured by the current socio-economic context. In fact, some of the greatest dangers faced by Mediterranean agrarian ecosystems come from the risk of large-scale land use changes such as abandonment or tree plantations. Due to the low economic competitive capacity of traditional exploitation, farming is economically dependent on external inputs (revenue support subsidies), among which agrarian subsidies are of prime importance. In this context of economic fragility of agriculture, there is an added problem with policy measures that may favour and/or accelerate land use changes incompatible with the present environmental benefits (i.e. afforestation reduces steppe bird population numbers). Liberalisation of trade would also most likely accelerate these changes, as the present exploitations are dependent on subsidies and far from being economically competitive.

5.   Countryside management in relation to agricultural land use

As seen in the preceding section, the fact of conformity with the evolution of the agricultural commodity market is not a sufficient basis for judging the opportunity of a particular agricultural land use. In addition to the environmental benefits reviewed above, certain agricultural land use practices have social benefits of various sorts, and these are associated with a broader consideration of how to manage the rural area of Europe. Some examples follow.

For a certain period, in particular since the 1950's, there has been an encouragement for agricultural structural measures to improve farmstead productive capacity. One of these has been drainage, both of waterlogged soils and of wetland areas (farm ponds, marshes, wetland areas along larger lakes). Recently, the awareness of the floral and ornithological value of water bodies has led to a more cautious approach to farm land structural measures. Saltmarsh grazing in the Fjand Enge area of the Nissum Fiord in Denmark encouraged the presence of large numbers of migratory birds - waders, ducks and geese - taking advantage of the sward maintained by the pastured livestock.

Particular flora associated with extensively grazed saltmarshes were also associated with such a land use. But the small field size discouraged the continuation of grazing, with the result of vegetation succession unfavourable for both the flora and the birdlife occurring on the abandoned parcels. Therefore a land reallocation scheme has been proposed for the consolidation of the saltmarsh holdings in order to make grazing financially viable. In exchange, certain farmers obtained access to soils suitable for intensive arable farming without drainage. The land with waterlogged soils has been maintained for summer grazing and wet-field haymaking [Nature and Forest Agency & Hansen, 1992].

The mountainous regions of Europe have poor soils and climate for arable land use, but are appropriate for extensive grazing. The market for livestock, however, does not by itself ensure a revenue related to extensive grazing that is sufficient to maintain such land use. The decline of such agricultural practice is not matched by the availability of substitute employment in mountain areas, as these also have a geographical handicap of access to major urban centres, which itself generates a second handicap in the availability of specialised services. Therefore there has been a phenomenon of social 'desertification' with the exodus of the rural population depending directly or indirectly on agriculture; the decline in their agricultural activity results in landscape changes which in turn make rural areas less attractive to other current or potential residents.

In order to palliate the lack of economic viability of mountain grazing, and linked to an overall policy for supporting other artisanal activities already existing in such regions but which also suffer from geographical handicaps (access to markets and services), comprehensive programmes have been initiated in France, in particular within the framework of parcs régionaux naturels. These combine headage payments under the Less Favoured Areas programme of the European Union (Map 3) with support for farmstead provision of tourist accommodation (gîtes ruraux), labels for regional produce, grant aid for the promotion of artisanal activity and local gastronomy, improvement of infrastructure (including systems of long-distance trails) and public services to encourage tourism (information distributed in major urban centres).

In Greece, the strategy for rural areas follows the same logic as that of maintaining a social fabric, with a particular objective of maintaining a cultural landscape that embodies not only traditional forms of agriculture (such as olive groves) but also regionally specific architecture. The abandonment of traditional agriculture specifically adapted to harsh environmental conditions results in a rapid deterioration of the landscape; in particular the development of scrub vegetation (leading to brush fires) and erosion - both resulting in the decrease of biodiversity. The decay of terraces used for orchards and arable agriculture diminishes the surface area for rainwater infiltration, leading to a local hydrological imbalance: desiccation impacting ground water resources and stream flow.

In alpine areas, summer meadows that are not grazed, develop high-grass swards, which fold beneath the snow in winter, providing a lessened frictional resistance to snow slides, resulting in an increase in potential avalanches. The consequences for human safety are two-fold: skiing accidents and destruction of housing. For this reason, special support is given to alpine shepherding in Switzerland. Throughout rural Europe, the maintenance of landscape features is an important accessory to making local areas attractive for the resident and tourist alike. Even minor features such as hedgerows and stone walls have been the object of public investment for their maintenance by the private landowner or farm tenant (otherwise these features are replaced by wire fencing), such as through the Countryside Stewardship scheme in the United Kingdom. The preservation of landscapes and support for the maintenance of landscape features was indeed one of the reasons for the setting up of National Parks in England and Wales, in areas where 'traditional' low intensity farming is essential for maintaining the character of the landscape.

Map 3. **Less favoured areas: Uplands**

Source: CEC DG-XVI, European upland areas, as defined by the EC.

6.   Policy measures in relationship to environmental benefits

Because of the diversity of both the agricultural practices and the variety of land use over large geographical areas, the policy measures to promote environmental benefits in agriculture have to be analysed on at least three territorial levels: the region, the nation and, when appropriate, the European Union. The types of measures in existence all reflect the recognition that agricultural land use is an inseparable component of countryside management. For this reason, in the application of a countryside policy, the local government authority or conservation agency will also be intricately involved in the elaboration of action to apply these measures. For the sake of simplification, only the national and European Union measures are examined here, as these are the framework for regional measures and local land use schemes.

Some of these are single-target measures, directed to ground water quality protection, landscape preservation or nature conservation. National land use planning systems specifically identify aquifer recharge areas and regulate land use so as to avoid impermeability of land cover and area wide as well as point source pollution. In some countries, such as Denmark, a wide variety of landscape elements are specifically protected (bogs, woodlands, stream banks and windbreaks), with even the possibility to establish covenants to bring land-development rights into the public domain without altering the basic status of private property holdings, rather than putting such elements into the public domain. These covenants may be indemnified through legal easements as under the Danish Nature Conservation Act, or be compensated for, through profit-foregone payments to compensate for agricultural field improvements or changes in products that are not allowed, as applied in Britain to protect Sites of Special Scientific Interest.

At the European Union level, two types of examples of the avoidance of negative effects (therefore having a beneficial result) are the Nitrates Directive, which limits the amount of nitrates which is allowed to permeate into ground water in areas under agricultural land use, and the Habitats and Birds Directives which require appropriate land use, agricultural or otherwise, or controls on recreational pursuits and natural resource use (hunting, picking or commercial trade in species), in order to maintain a favourable conservation status of certain species of flora and fauna. Certainly the first of these measures requires compliance to compulsory legislation according to the polluter-pays-principle.

Other means are multiple-target measures. At the national level, two examples already mentioned (the Danish Nature Conservation Act and the British Countryside Stewardship scheme), are intended to promote a balanced approach to countryside management. Within the European Union, the Less Favoured Areas regulation, directed to the mountain areas of European Union Member States, has already been given as an example. There are also the structural objectives, in which large zones are identified for specific financial aids in order to achieve economic development, complementary to the strong economic evolution of other regions, in order to avoid structural imbalance in the European economy. Many aspects of such development aid have an influence on both agricultural land use and environmental quality together. This European Union policy is also reflected through the possibility for aid programmes to assist non-EU states to implement projects having an environmental component (PHARE, TACIS and the Danube Basin Programme), and again such projects can influence agriculture and the environment together.

Within every country there are measures specifically intended for agricultural land use as such, and some have a direct influence in protecting agricultural land uses that are directly associated with landscape values, such as the control of vineyards in France through the system of appellation d'origine controlée. Within the European Union the promotion of environmental benefits through agricultural land use is harmonized to some extent by a series of measure, presented in Table 2.

Table 2. European Union policy instruments in favour of the environment and related to agricultural land use practices under the Treaty of Rome, as modified by the Single European Act

Article 130r	*Relating to environmental responsibility.*
Directive 79/403 "birds"	Identification of bird species of European importance, with measures relating to the protection of their habitats.
Directive 91/676 "nitrates"	Voluntary and mandatory schemes to protect ground water quality through seasonal restriction in fertilizer use, storage of livestock manure, limitation of fertilizer application charge according to soil characteristics, climate and nitrogen requirement of crops.
Directive 43/92 "habitats"	Provides for the protection of habitats and species (other than birds, already provided for through Directive 79/403) of European importance, establishes the principle of a European ecological network.
Article 42	*Relating to agricultural policy.*
Regulation 2052/88 Structural Funds, Objective 1 & 5B areas (Map 4) (including "less favoured areas" provisions)	Support for farm improvement (irrigation) and product diversification, farm-based tourism, environmental protection measures; headage payment to maintain livestock farming in "less favoured areas" with environmental handicaps in terms natural production potential.
Regulation 2092/91 "organic agriculture"	Applies to unprocessed agricultural products, to products intended for human consumption composed essentially of one or more ingredients of plant origin, and it introduces specific rules for the production, inspection and labelling of such products.
Regulation 1765/92	Market control requirement for set-aside on farms producing more than 92 tonnes of net cereal, oilseed and protein-rich crops; introduces a beef premium and measures to control livestock grazing density.
Regulation 2078/92 (including "environmentally sensitive areas)	A series of environmental measures, including long-term set aside (twenty years) and compensation for landscape and habitat management through maintaining the effects of "traditional" agricultural land use practices.
Regulation 2080/92	Encouragement for woodlot planting and management; preferential funding with regard to deciduous varieties when locally appropriate.

Map 4. **Objective 1 and 5(b) areas**

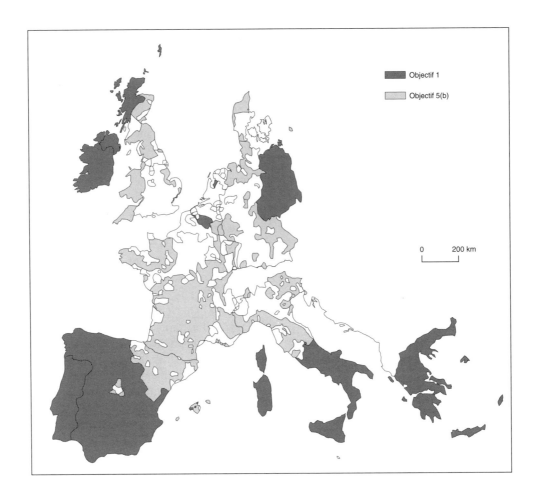

Source: CEC, Areas eligible for EC Objective 1 and 5(b) Structural Funds.

7. Environmental influence of agriculture compared to other land uses

One way to place the environmental influence of agriculture into perspective when compared to other land uses is through the market value of agricultural land. On the one hand, the environmental factors suitable for highly productive agricultural land, such as the well drained alluvial soils, are also similar for other land use (urban development, infrastructure and forestry). On the other hand, "parks and nature preserves are located in the unproductive uplands, or on dry, infertile, hilly, or rocky sites that are unsuitable for intensive agriculture" [Huston, 1996]. The competition for land without environmental constraints will in any case mean that maintaining any environmental benefits associated with open land are more probable with regard to agriculture (or forestry) than other land uses. The presence of environmental constraints having an influence on agriculture also work in favour of avoiding the deterioration of biodiversity and landscape values associated with non-agricultural land use, with the exception of outdoor recreation such as skiing, golf, hunting and cross-country motoring (even in these cases, it is a matter of how the provision for such sports is made: environmental damage is always a question of degree).

"The fortunate side of this situation is that the economics of land conversion have allowed the preservation of most of the biodiversity on these landscapes. The landscape spatial pattern component of biological diversity is preserved by the large areas of low productivity land that are set aside and maintained with a very low density of human population. The species diversity component is preserved by the inherently high within-habitat diversity of many of the low (not necessarily the lowest) productivity environments, as well as potentially high between-habitat and geographical diversity." [Huston, *op. cit.*]

To resume, agricultural land use will be in danger of displacement by other land uses sharing similar advantages from environmental suitability or having a better aptitude to environmental constraints. In both cases, in order to preserve the environmental benefits accruing to agricultural land use, it is necessary to influence the market value of land. Some notable examples where intervention could provide useful benefits for the environment would be to avoid the plantation of pine forests on grazing land and of eucalyptus groves on arable land.

8.   The effects of trade liberalisation in the European context

In order to assess the effects of trade liberalisation on the different types of environmental benefits which can be associated with agriculture, the work by ECNC leads to the following observations. An assumption exists that trade liberalisation in the European context will encourage a rational use of land resources. The question remains as to whether the market will direct agricultural land use to the most appropriate locations, with the use of the most appropriate technology, from the environmental point of view. With regard to actual policy to regulate or orient agricultural land use, many examples have been given in the preceding discussion which demonstrate that an imperfect market exists and that externalities such as air, soil and water quality, landscape features, and the social welfare of rural areas are not necessarily dealt with efficiently or fairly in the market. For these factors, can their apparent externality be internalised?

According to the precautionary principle, special care should be accorded to the use of non-renewable resources, and by extension to those resources which are regenerated only after a long period of time. Environmental benefits are therefore not to be hypothecated lightly, and the effects of trade liberalisation on agricultural land use is thought to entail certain dangers in particular for biodiversity and landscape values in Europe. One of these dangers is thought to be the rationalisation of agricultural practice through standardized production methods and their implementation in larger farm units in order to achieve economies of scale. Land market values will certainly be a factor in such a trend. But, on the other hand, a more efficient use of inputs and a search for the most appropriate mechanical support can attenuate potentially negative impacts on air, soil and water quality.

A second of these dangers is thought to be the marginalisation, or even abandonment of agricultural land. Here the risk is probably real enough, for land market values, even if low, will not drive an economic system; only demand will. The development of specialised commodity markets, such as for free-range meat, may palliate a lack of demand. How elastic such demand will be remains to be seen, and it is not certain that such commodities would provide sufficient employment in order to support a rural social fabric, if agriculture is indeed the critical sector.

The third danger is therefore with regard to countryside development: will there be enough economic vitality in rural areas to maintain an agricultural population outside of a restricted number of small towns and large villages. Programmes to supplement (agri-environmental measures) and diversify (green tourism measures) agricultural revenues may prove to be sufficient. But only the renewal of the current agricultural population in the succeeding generations will provide the answer.

In every scenario presented above, a premise is that trade liberalisation will perhaps reduce agricultural activity in Europe. Whether the world market will change radically in the coming years is, however, open to debate. Climate change, for the sake of example, could easily alter present policy positions with regard to food security, if not with regard to humanitarian aid. Market reality is more about food than environmental benefits.

One conclusion is that some environmental benefits are possibly better assured in a context of liberalised trade as compared to a high-price policy; certainly these are the ones that most directly make an impact on human society and the perennity of agriculture itself: air, soil and water quality. A second conclusion is that some environmental benefits may not be assured in such a context, and these have to do with the very substance of what makes Europe in many ways unique: the biodiversity and landscape values of cultural landscapes, maintained only through a living social fabric in rural areas, nourished by agricultural activity. To be fair, measures in favour of these latter environmental benefits may not succeed in the long term, market or no market. They require human commitment, effort and knowledge; none of these most important of all resources can be guaranteed.

9. Conclusions regarding the environmental benefits of agriculture in Europe

Several simple general conclusions can be made through this overview of the environmental benefits of agriculture in Europe. First, present policies are based on the principle of joint-production of commodities and amenities. With proper signals, agriculture should be able to provide both food and a rich array of environmental benefits. Second, these policies are also based on the premise that compensation is required for the non-market provision of environmental benefits, outside of the context of the polluter-pays-principle. It is admitted that there are cases where desired environmental benefits are not inherent in the current forms of agricultural land use itself. Third, that an appropriate framework for trade liberalisation is a recognition that the market is imperfect, and that the precautionary principle requires a prudent approach to handling the economic future of certain parts of the agricultural economy which are the most directly associated with the provision of the more fragile environmental benefits stemming from agriculture over its long history in Europe: biodiversity and cultural landscapes.

10. Bibliography

BARNETT, V., R. Payne and R. Steiner eds. (1995), *Agricultural Sustainability - Economic, Environmental and Statistical Considerations*, John Wiley & Sons.

BEAUFOY, G., D. Baldock and J. Clark (1994), *The Nature of Farming - Low Intensity Farming Systems in Nine European Countries*, Institute for European Environmental Policy.

BETHE, F. and E. Bolsius, eds. (1995), *Marginalisation of agricultural land in the Netherlands, Denmark and Germany*, Ministry of Housing, Spatial Planning and the Environment, the Netherlands.

Commission of the European Communities (1991). *CORINE Biotopes*, CEC.

Eurostat, 1995: *Statistical Compendium for the Dobrís Assessment - Europe's Environment*, Office for the Official Publication of the European Communities.

HALLADAY, P. and D. Gilmour (1995), *Conserving Biodiversity Outside Protected Areas: The Role of Traditional Agro-Ecosystems*, IUCN.

HOGGART, K., H. Buller and R. Black (1995), *Rural Europe - Identity and Change*, Arnold.

HUSTON, M. A. *Biological Diversity - The Coexistence of Species on Changing Landscapes*, Cambridge University Press.

JÄGER, J., A. Liberatore and K. Grundlach eds. (1995), *Global Environmental Change and Sustainable Development in Europe*, Office for Official Publications of the European Communities.

JORDAN, T.G., *The European Cultural Area*, Harper Collins.

National Forest and Nature Agency, Hansen J.M. *Nature Management in Denmark*, The Ministry of the Environment, The National Forest and Nature Agency, Denmark.

McCRAKEN, D.I. and E.M. Bignal eds. (1995), *Farming on the Edge: The Nature of Traditional Farmland in Europe*, Joint Nature Conservation Committee, United Kingdom.

NOWICKI, P.L. (1995), *Nature in Trust*, Biological Journal of the Linnean Society, vol. 56.

POUNDS, N.J.G. (1990), *An Historical Geography of Europe*, Cambridge University Press.

STANNERS, D. and P. Bourdeau eds. (1995), *Europe's Environment - The Dobrís Assessment*, European Environment Agency.

TIVY, J. (1990), *Agricultural Ecology*, Longman.

TRACY, M. (1993), *Food and Agriculture in an Market Economy*, Agricultural Policy Studies.

VAN DER WEIJDEN, W. Lof H. and J. Warner (1992), *EC Agricultural Policy and the Environment*, CLM (Centre for Agriculture and the Environment, Utrecht).

Annex: Biodiversity Assessment for Identifying Environmental Benefits of European Agricultural Land Use by D.M. Wascher, ECNC

1.   New methodologies and approaches

The European Commission's "Fifth Environmental Action Programme ("Towards Sustainability") has been developed as part of the Community's preparation for the Earth Summit of Rio de Janeiro and marks Europe's growing awareness of the importance which sustainable principles have for both the environment and the economic market system. Reviews of this programme (EU-Commission conferences in 1994 "Towards a new development approach" and in 1996 "New European Policies for 1996-2000", Brussels 15-16 February 1996) as well as environmental reports such as the EEA's "Europe's Environment - The Dobris Assessment" [Stanners and Bourdeau, 1995] recognise the urgent need for new methodologies and approaches in order to assess the functional links between economic processes and biodiversity.

This holds especially true for the large portion of agricultural areas within Europe's rural environment which are known to provide numerous actual and future environmental benefits for biodiversity, agriculture, tourism and recreation as well as human health. On a continent where the former natural vegetation has been changed almost entirely under the influence of human land use activities, large proportions of the existing ecological and socio-economic values are more or less directly dependent on the way the land is being managed. While European nature conservation legislation focuses on some of the most important key-areas for rare and endangered species and habitats, the remaining large areas - dominated by agriculture and forestry - are increasingly facing changes in management regimes that devalue both their long-term economic potential and environmental functions important for Europe's nature and wildlife.

"One unfortunate consequence of the economic regulation of land use is that all the productive ecosystems in the world have been destroyed and converted to agriculture. In case after case around the world, the productive lowland and alluvial areas have been almost completely converted to agriculture and the indigenous flora and fauna nearly eliminated, while the parks and nature reserves are located in the unproductive uplands, or on fry, infertile, hilly, or rocky sites that are unsuitable for intensive agriculture. The fortunate side is that the economics of land conversion have allowed the preservation of most of the biological diversity of these landscapes. In conclusion, the fact that agricultural productivity and biological diversity are strongly influenced by the same environmental conditions should provide an economic rationale for the preservation of biological diversity". [Huston, 1994].

The knowledge of how environmental legislation and economic steering mechanisms are actually affecting the ecological values of Europe's agricultural lands is limited to individual case studies which are able to prove functional links between the economic incentives, the choice of land management/production and the resulting effects on the local fauna and flora. While these examples are generally well documented and scientifically stable, the overall implications remain rather diffuse and lack concrete policy implications. The reason can be seen in the following facts:

- there is a lack of comprehensive, harmonised information of the distribution and status of biodiversity values for Europe's habitats and landscapes;

- reliable quantitative and qualitative information on how the EU-subsidies are precisely affecting these biodiversity values throughout the Union do not exist;

- current research programmes on sustainable aspects of agricultural land use practices are focusing primarily on energy and nutrient cycles of specific farmsteads while excluding or simplifying biodiversity aspects such as habitat functions and species viability;

- environmental reports are more precise in focusing on the results of the environmental changes (numbers of threatened species, habitat fragmentation, etc.), than on their causes (human activities, economic mechanisms);

- there is a lack of officially acknowledged environmental indicators that meet both scientific and policy standards;

- the geographical, climatic, ecological and cultural conditions of Europe are extremely diverse and require a high level of biogeographic specifications, largely ignored by current European Union procedures.

For improving the quality of environmental information related to biodiversity currently found in agricultural ecosystems, data on human land use and on nature is of special interest. Only a few of these sources are sufficiently harmonised to allow European-wide assessment procedures. Up to the present time, however, the available capacities of interpreting and manipulating remote sensing data have only been partially exploited. Remote sensing data exists on some country levels, for parts of the European Union territory (CORINE land cover) and for pan-Europe (e.g. at ESA). Precise geographic identifications of land use structures, landscape elements, landforms, patch sizes and socio-economic patterns provide a wealth of information that can be linked to other layers such as topography, climate, soil and vegetation data. While remote sensing data does not cover all information needs, it should be considered as an important pre-requisite for any subsequent assessment strategy.

2.   Pilot studies for assessing Europe's biodiversity

In the light of the first work programme of the newly established European Environment Agency [Copenhagen 1994], the European Centre for Nature Conservation (ECNC) has been asked to lead pilot studies to develop a methodology for assessing the state and trends of biodiversity in Europe as part of the work programme of the European Topic Centre on Nature Conservation (Paris). The ten participating national expert centres on nature conservation and ecology, performed different forms of 'medium filter' approaches (Box 2) to report about land use, habitats, landscapes and species within selected areas of their biogeographic region.

The pilot studies in ten biogeographic regions covered an area of 16 115 km². A first phase was conceptualised as a 'medium filter approach' to perform a biodiversity assessment based on top-down defined assessment criteria in a standardised format, resulting in a total of about 5 000 individual records on species, habitat connectivity, human activities and legal issues of nature conservation. A second phase - designed as a bottom-up approach involving expert knowledge at the regional level - focused on *agricultural* habitat types within the same study areas and analysed qualitative and quantitative aspects of biodiversity for the assessment categories 'naturalness', 'ecological quality', 'habitat threat', 'species value' and 'landscape value'.

Both phases of the pilot studies have demonstrated that biodiversity values could be recognised in large proportions of the study areas, clearly exceeding the area of actual (legal) nature protection and frequently comprising wide ranges of semi-natural or cultural landscapes. The observed biodiversity values included: (i) presence of endangered or rare species; (ii) high species richness per area unit; (iii) well developed and maintained habitat connectivity; (iv) high structural diversity; and (v) highly suitable forms of agricultural land use (depending on soil quality or wildlife management objectives).

These findings indicate that current approaches in evaluating the beneficial role of agricultural land use need to be broadened with regard to specific habitat conditions as an integrated part of the landscape dimension. There is a very obvious need for primarily two instruments: (i) standardised assessment tools for evaluating habitat quality at the site level; and (ii) geo-referenced baseline data for evaluating adequate habitat distribution at the landscape level.

3.   Habitat quality

Following the example of ITE's Biotopes Occupancy Database (BOD) [Eversham, 1993] the proposed future methodology is designed to develop 'habitat profiles' for natural as well as semi-natural habitat types. Based on the coding system of the Palaearctic Habitat Classification System [Devillers & Devillers-Terschuren, 1993], such a reference list is supposed to provide information on the presence of typical plant and animal species, and physical as well as management aspects, based on the following criteria: (i) complete list of plant species (trees, shrubs, herbal layer, lower plants) characteristic for the habitat type; if necessary differentiating eco-regional sub-types within the same hierarchic level; (ii) minimum set of species that are indicative for habitat quality or pressures; (iii) individual species that are indicative for habitat quality or pressures; (iv) list of characteristic species from relevant animal species groups; (v) characteristically appearing exotic or invasive species; (vi) adequate land use for maintaining the habitat; (vii) typical soil, topography, and hydrological conditions. The resulting habitat profile reference base will provide the measuring scale when monitoring habitat quality at the site level. The results of the monitoring will feed back into the Habitat Profile Reference Base.

Box 2. The terms "coarse and fine filter" analysis

The terms "coarse and fine filter" analysis derive from American strategic approaches to ecological assessment, developed in the eighties [Noss 1983 and 1987]. Recognising the link between species and habitats as well as between vegetation type diversity and high edaphic variety or topographic relief, the analysis of landscape-sized samples allows the prediction of species presence depending on specific or multiple habitat needs.

However, wide ranging species or species with very local or restricted distributions are not likely to be captured by this "coarse filter" approach and require additional "fine filter" tools. While originally used as a strategic approach in safeguarding species through habitat conservation, the pilot study report [ETC/NC, 1996] uses these terms in a wider sense to characterise differences in the scope, resolution and objectives of biodiversity assessment techniques.

The report therefore also introduces the concept of a "medium filter" approach to describe the intermediate role of the Pilot Study approach.

The requirement to implement pilot studies for testing a methodology for the assessment of the state and trends of biodiversity in Europe pointed at the following options:

- a "fine filter" approach (sampling techniques) to gather field data systematically throughout Europe on species, habitats and human activities, taking into account seasonal aspects and increased staff requirements; the establishment of repeatable assessment procedures when gathering and comparing data over longer periods of time to detect changes;

- a "medium filter" - approach to assess biodiversity on the base of existing data capacities at the national and regional level;

- a "coarse filter" - approach as the initial step based on existing data at the European level.

---

4.   Habitat representation

The need for reliable baseline data for 'adequate' habitat representation at the landscape level can derive from multiple sources, including data on the potential natural vegetation, soil and land cover. One other important tool for identifying minimum habitat size and distribution will be the establishment of species profiles. The profile of species which are characteristic for certain habitat types - e.g. the white stork for lowland wet grasslands - provides useful indications for adequate habitat requirements to ensure 'favourable conservation status'. For the example of a landscape with fragments of lowland wet grasslands, it is possible to compare actual distribution and population of the white stork with its potential distribution. Such assessments would not be limited to one habitat and land use type only, but encompass a range of different options within the biogeographic range of the landscape.

These concepts point at the need for integrated assessment procedures that take into account the spatial and functional relations between human land use and habitat as well as species diversity. On the base of remote sensing information and ground validation, land use evaluation procedures and long-term monitoring programmes can be systematically developed, applying additional digitally geo-referenced and statistical information as well as selected environmental indicators.

Because of Europe's geographic and socio-economic diversity, the development of a methodology for assessing the state and trends of biodiversity in Europe is supported by a project on identifying ecological regions as the reference base for future assessment procedures. Ecological regions have already been developed in many countries, however only in a few cases, e.g. Canada and Great Britain, are they also actively used for assessment and management purposes. The reported results are very encouraging: the use of ecological regions has led to new and stronger links between local populations and their administrations and regional environmental characteristics and assets [English Nature, 1994; Nowicki *et al.*, 1996]. This is primarily important for the development and introduction of sustainable land use principles.

5. Conclusions

The pilot studies have demonstrated opportunities to apply 'coarse filter' analysis modules for identifying centres of potential high and moderate biodiversity and to establish coarse baseline data as a reference for detecting changes. The 'coarse filter' approach is to be followed by 'fine filter-modules in the form of site and area monitoring activities at the site level [Natura 2000], the ecosystem level and at the landscape level. The process will have to contribute to improving the coverage, actualness and detail on semi-natural habitats, by using and improving remote sensing data and by integrating socio-economic information into monitoring assessment procedures.

For this purpose, ecological regions seem to be the logical spatial unit to determine sustainable forms of land management that take into account the adequate representation and integration of biodiversity properties such as semi-natural habitats, habitat connectivity in the form of corridors and buffer zones, as well as species richness and rarity. The studies have also identified the need to further improve the methodological tools for assessing semi-natural and cultural landscape features and to strengthen the inter-disciplinary co-operation between nature conservation agencies and agricultural department at the level of research and administration.

6. Bibliography

Commission of the European Communities (CEC) 1992. Towards Sustainability: a European Community programme of policy and action in relation to the environment and sustainable development; Brussels, Commission of the European Communities).

Council of Europe, 1987: *Management of Europe's Natural Heritage - Twenty-five years of activity*; Environmental Protection and Management Division, Council of Europe, Strasbourg.

BEAUFOY, G., D. Baldock and J. Clark (1994), *The Nature of Farming - Low Intensity Farming Systems in Nine European Countries*, Institute for European Environmental Policy.

DEVILLERS, P. and J. Devillers-Terschuren (1993), *A classification of Palaearctic habitats*. Report to the Council of Europe.

English Nature, (1994), Keith Duff, English Nature, ECNC seminar 10 November 1995.

ETC/NC (1996), *Methodology for an Assessment of Europe's Biodiversity*. MN2.1 Asub-Project Dinal Draft, submitted to the European Environment Agency by the European Topic Centre on Nature Conservation, based on a draft report by D.M. Wascher, European Centre for Nature Conservation, Tilburg, The Netherlands, 32 pages.

EVERSHAM, B.C. (1993), *Biogeographic research in the Biological Records Centre*. ITE Annual Report 1992-3, 22-25.

FAITH, D.F. (1995), *Biodiversity and Regional Sustainability Analysis*. Division of Wildlife and Ecology, CSIRO Australia.

HUSTON, M.A. (1994), *Biological Diversity*. The coexistence of species on changing landscapes. Cambridge University Press, Cambridge, 681 pages.

JESINGHAUS, J. (1995), The Pressure Indices Project: Theory and Structure. Commission of the European Union, EUROSTAT, unpublished draft, 42 pages.

JONGMAN, R. (1995), CORRIDOR: An assessment of functional ecological corridors for the conservation of biodiversity in a landscape framework, ECNC Tilburg (unpublished).

NOWICKI, P., G. Bennett, D. Middleton, D. Rietjes and R. Wolters eds. (1996), *Perspectives on Ecological Networks*, European Centre for Nature Conservation.

STANNERS, D. and P. Bourdeau (1995), "Europe's Environment - The Dobris Assessment". The European Environment Agency, Copenhagen, 776 pages.

ENVIRONMENTAL BENEFITS OF AGRICULTURE: NON-EUROPEAN OECD COUNTRIES

by
Sandra S. Batie[11]
Michigan State University, USA

Agriculture can be thought of as an argument with nature. The farmer's pursuit of production of few species of plants or animals runs counter to the natural tendency toward more diversity. Natural environments tend to have diverse species and low rates of nutrient cycling as well as resilience to weather or disease events [Soule and Piper, 1992]. Conventional agriculture, with its emphasis on high yielding monocultures achieved with additions of purchased inputs, frequently manifests high rates of nutrient cycling. In addition, conventional agriculture can evidence low resilience to weather or disease events. By its very nature, cultivation, disrupts the natural environment, as does livestock production that exceeds the carrying capacity of the available land.

However, it is clear that agriculture can be complementary to natural processes. For example, in addition to providing food, fibre, and rural employment, agriculture can provide habitat for wildlife; it can build soil quality; it can protect water and air quality; it can recycle nutrients; it can control flooding; it can protect genetic material; and, it can provide appealing landscapes that in turn provide recreation and tourism opportunities[12].

An important question emerges: how can agriculture enterprises reduce their negative impacts on the environment and enhance the positive ones? The answering of this question, however, presumes an agreement on what a desirable environment entails - as definitions of desirable environmental attributes vary across and within societies. These definitions also change over time.

Furthermore, agriculture operates within an institutional context that includes agricultural and trade policies. To what extent can reform of these policies lead to an enhanced environment? What is the role of market forces in influencing the environmental impacts of agriculture? Does the enhancement of the environment require the adoption of more sustainable systems of production? This paper addresses these issues with particular attention to examples from the non-European OECD countries.

---

[11] Sandra S. Batie is the Elton R. Smith Professor of Food and Agricultural Policy, Department of Agricultural Economics, Michigan State University, East Lansing, Michigan, USA 48824-1039. Ph 517-355-4705, Fax 517-432-1800. Email: batie@pilot.msu.edu.
[12] Bromley (1996) notes that these agro-environmental benefits can be thought of as belonging to one of three classifications: amenities, habitats, or ecological processes.

1. Desirable environment: Definitions

Cultures, as well as subcultures, will differ as to the definition of a desirable environment. North Americans, Australians, and New Zealanders, for example, tend to value an ecosystem that is "natural," that is, more like their perceptions of the wilderness that would have existed without humankind's involvement[13]. Asian cultures are less likely to define a desirable environment as synonymous with wilderness and more frequently view the natural environment as providing raw materials for economic development purposes. Both European and Asian cultures may highly value a managed environment such as an open and varied agrarian landscapes or formal gardens [Rayner, 1991; Vail, 1995]. Even the definition of a desirable species may vary - wolves, rabbits, geese and deer may be considered beneficial in one region; but unwanted in others [OECD, 1995].

Some cultures will place a high value on protecting wildlife habitat, genetic diversity, or unpolluted ecosystems; still others on protecting human health or rural lifestyles. For example, within Western cultures health issues are only one dimension of a desirable environment. In contrast, in Japan, the central preoccupation of environmental policies has been the threat posed by pollution to human health [Vogel, 1990]. Mexicans are concerned about the threat posed by pollution of water to both human health and agricultural productivity; water quantity issues are also of considerable concern in Mexico.

These broad generalisations as to the definitions of a desirable environment obscure cultural divisions within a nation [Edgell and Nowell, 1989; Rayner, 1991]. The many definitions of a desirable environment that are found within a nation. These are often reflected in various political arguments surrounding proposed public policies. In addition, different cultural interpretations of scientific and technical knowledge of agricultural impacts on environmental health may be reflected in different societal assessments of the benefits, costs, or risks associated with impacts on natural systems such as the use of agricultural chemicals.

Values also change over time. Historically, both Western and Eastern cultures have undergone profound changes in attitudes toward the natural environment. These attitudes are also reflected in customs and institutions such as property rights [Ittelson, Proshansky, Rivlin, Winkel and Dempsey, 1974]. Hays (1958), for example, documents how environmental values in the United States have changed from a dominant theme of conservation prior to World War II to that of environmentalism after World War II. Conservation emphasizes the more efficient use and development of physical resources as well as the protection of the physical productivity of resources. Environmentalism focuses on reducing pollution, protecting ecosystem functioning, and developing appropriate scale technologies. Environmentalists advocate, for example, that forests should be considered less as commodities and more as an environmental ecosystems with high amenity, recreational, and wildlife values. As the conservation ethic yields to environmentalism in the United States, concern for the protection of on-farm productivity through the conservation of soil resources is shifting to concern for off-farm impacts of agricultural systems on the environment.

Because the objective of a desirable environment presupposes an acceptable definition of the characteristics of such an environment, and because such definitions vary over cultures, subcultures, and over time, it is not surprising that public policies in pursuit of desirable environments also vary.

---

[13] For a good discussion of wilderness in Canada, Australia, and the United States, see the collection of articles in the Winter 1989 *Natural Resources Journal* 29(1). The collection is called "Wilderness: Past, Present and Future".

Agriculture as a provider of services

A useful definition of a desirable environment also requires a better understanding of agriculture's relationship to the environment as a provider of services. It is informative to characterise agriculture as providing a set of services. As Figure 1 illustrates, these services can be positive, that is, they can be "benefits"; or, these services can be negative, that is, they can be "disservices," also referred to as "costs" or "damages".

Figure 1. The environmental services of agriculture

A particular agro-environmental service may fall on a continuum that stretches from "extremely damaging" to "highly beneficial". However, an assessment of any particular service will depend on who is doing the assessment and what is the assumed environmental or social goal.

Clarifying agriculture's relationship to the environment requires precise definitions of these services. That is, it is more instructive to identify the service of "agricultural land transforming animal wastes into valuable soil conditioners" than it is to refer to "agriculture protecting soil quality". Two other examples are: (i) "grain farming providing fall feeding for migrating Canadian geese" rather than "agriculture providing wildlife habitat"; and, (ii) "agricultural hedgerows providing an appealing landscape feature" rather than "agriculture providing appealing landscapes".

Precise definitions of services illuminate the situations when services are in conflict with one another. For example, to obtain more, say, English sparrows, by the management of agricultural systems, one must forego another species' success such as the bluebird, as sparrows tend to out compete bluebirds when they occupy the same area. Or, to obtain more water quality through reduced use of chemicals may require reductions in yields of certain crops. Or, to obtain more wilderness species such as the American black bear will mean the loss of agricultural dependent species such as the opossum. Or, to obtain more surface water quality with conservation tillage may diminish groundwater quality or reduce available fodder for livestock.

There are cases where services complement one another. An example is where improved soil quality obtained by crop rotations improves yields and provides a buffering effect that protects water from eutrophication[14]. Another example is where amenities contribute to rural development [OECD, 1994], such as the tourism to visit traditional farming areas.

---

[14] As Rausser and Foster (1991) note, these conflicts can be though of as joint production processes. That is, many of the inputs used in agriculture (such as pesticides) produce both valuable and undesirable services. Rausser and Foster also note that this jointness can be found in consumption. That is, consumption of food can provide the "goods" of nutrition and taste, but it may also provide "bads" such as pesticide residues that may adversely affect human health.

Variations over time

The services provided by agriculture vary overtime. Fitzpatrick (1995) provides an interesting example of the change in services overtime with reference to barn owl habitat; a species whose population increased dramatically with the arrival of agriculture to the American Midwest. However, the change in mid-west agriculture from dependence on forages in the 1800s to the present day dependence on row crops led to the decline of the once common barn owl. Fitzpatrick notes: "Most habitat changes taken individually are small, involving only a portion of a field or farm. But however slight the separate changes, they may accumulate over time, particularly if the change is a trend among farmers in an area. Changes in land management can either improve or degrade the habitat's value for raptors; to maintain or increase raptor populations it therefore is necessary to identify and implement those changes that favour them"[15]. There are many similar stories about the changes in the type of services provided by agriculture over time.

Services and agricultural practices

There are agricultural practices as well as agricultural systems that either mitigate the harmful effects of agriculture on the environment or enhance the beneficial ones. Most research has focused on those practices and systems that mitigate potentially harmful environmental effects of agriculture. For example, after a comprehensive literature review of agricultural practices and systems that reduce agriculture's off-farm negative impacts, a U.S. National Academy of Science study committee [National Research Council, 1993] identified four major types to be the most effective[16]: (i) those that conserve and enhance soil quality; (ii) those that increase nutrient, pesticide, and irrigation use efficiencies; (iii) those that increase the resistance of farming systems to erosion and runoff; and (iv) those that use field and landscape buffer zones.

Within these four sets, there are many examples. For example, rotation of crops can conserve and enhance soil quality. If done appropriately, rotational grazing[17] can distribute manure as part of the natural cycling process; it can reduce the need to either purchase feed or to bring self-harvested crops to confined cattle; and, it can reduce erosion and runoff by eliminating confined feedlots [Murphy, *et al.*, 1996; Washburn, *et al.*, 1996; Petrucci, 1996]. Filter strips and field scale buffer zones, if large enough, are other examples of effective technologies that reduce off-farm impacts, as is conservation tillage [National Research Council, 1993]. Reducing livestock stocking rates to protect grass cover can reduce erosion. Rice paddy fields in upland areas can reduce water runoff, prevent flooding in lowlands and limit soil erosion [Yoichi, 1992]. Biologically based pest controls, particularly if embedded in an integrated management program can reduce the need for pesticides [National Research Council, 1993; Office of Technology Assessment, 1995]. There are also existing

---

[15] Fitzpatrick (1995) notes that changes to provide raptor habitat do not necessarily require the abandonment of row crop agriculture and a return to forages. Rather, protection needs to be provided for raptors' favorite food, voles, via a grassy non- cultivated strip; raptor hunting perches need to be provided where trees are not available. However improving raptor habitat may reduce other species than voles, such as pheasants (because raptors will eat pheasant chicks).

[16] The National Research Council (1993) advocated that policies designed to obtain wide scale adoption of these practices, using public funds, should be targeted to problem areas such as valuable watersheds. Furthermore, within those problem areas the practices should be further targeted to problem farms that account for most of the off site damages.

[17] Rotational grazing involves the farmer moving cattle between heavily grazed fields to heavily foraged fields and then allowing the heavily grazed field to reestablish.

and foreseeable biotechnology products that provide resistance to weeds, insects or diseases [Office of Technology Assessment, 1995].

The effectiveness of any particular practice or system will vary by the topography of the farm or the ecosystem in which the farm is located. A "one size solution" will not fit all problem areas or farms. Conservation tillage, for example, is better suited for well-drained soils in moderate climates [National Research Council, 1993]. It is less likely to be adopted where the use of residues is important to feed livestock, such as is the case in some of Mexico's semiarid regions [Erenstein, 1995]. Filter strips are suitable for wetter climatic regions.

The National Research Council's list focused mainly on those practices that mitigate the off-farm negative impacts of agriculture. There are also many examples of managing a farm so as to enhance the values of certain agro-environmental services for wildlife, for landscape appeal, for nutrient cycling or for tourism. These practices can include the provision of wildlife habitat by leaving acreage un-tilled, by planting of favoured food sources for specific wildlife, by delaying harvest until after bird nesting, by maintaining buildings or stone walls of historical or scenic interest, by appropriate spreading of manures, or by protecting forests, meadows, and woodlots.

Many of these practices - either to mitigate negative impacts or to enhance positive ones - do not increase profits, and farmers do not readily adopt them without additional incentives. There are many examples of such incentives, however. Along the Blue Ridge Mountains scenic highway in the Eastern United States, farmers are paid for a "scenic easement" that pays a set amount on the condition that farmers maintain historic buildings and split rail fences within sight of the highway. Japan uses grants to local governments to promote more environmentally friendly low input agriculture [OECD, 1995]. Australia has a National Landcare Programme that provides grant funds to landcare groups to enhance the sustainable use of land, and Canada's Green Plan provides funds to assist farmers in developing better land management practices [OECD, 1995].

In other circumstances farmers are required, without compensation, to adopt certain practices or farmers are freed from certain liability or nuisance suits if they adopt certain practices. Japan, for example uses strict land use controls to control erosion: New Zealand's local officials have the authority to enforce environmental standards and to prohibit unsustainable land use practices [OECD, 1995]. The United States federal and state laws limit conversion of ecologically significant wetlands to farming uses, as well as regulating the discharge of large volumes of manure into waterways.

It is not true, however, that all agro-environmental oriented practices are unprofitable. There are practices and systems referred to as "complementary technologies" that, if used appropriately, can provide both profits and positive agro-environmental services [Office of Technology Assessment, 1995]. Many authors refer to these complementary technologies as "win-win". That is, a "win-win" technology is one which the farmer views as profitable and which produces a positive environmental benefits. An example might be a nutrient management plan that both reduces the purchase and use of commercial fertilizers, but does not negatively impact yields [National Research Council, 1993]. The set of "complementary technologies" is expanding as researchers devote more attention to agro-environmental problems; precision farming in particular, has considerable potential in this regard. "Win-win," or complementary technologies, should not require permanent incentives for adoption; however, start-up incentives or educational programs may be necessary [Batie, 1994]. In some countries, associations, often farmer-led, are voluntarily implementing strategies to improve local environments [OECD, 1996].

2. The role of sustainable agriculture

In some cases, the agricultural technologies and systems which mitigate agro-environmental problems or which enhance environmental quality are classified as belonging to sustainable agriculture[18]. Although sometimes the concept of sustainable agriculture incorporates a philosophy of being in "harmony with nature" (as opposed to the more conventional "management of nature") [Batie and Taylor, 1991], it more frequently refers to alternative practices or alternative systems regardless of underlying philosophies. Indeed many farms employ neither totally conventional nor sustainable systems, but are more-or-less "conventional" or "sustainable" in their main tendencies [General Accounting Office, 1991]. Farmers who adopt reduced input systems frequently do so to solve a particular production, environmental, or health problem as opposed to doing so for philosophical or ideological reasons [Buttel, *et al.*, 1986].

Sustainable agricultural practices are so-named because, if used properly, they are more environmentally protecting than conventional agricultural practices. While organic practices are almost always considered sustainable, since they use no purchased chemical inputs, there are many sustainable practices, that are low-chemical but not organic. These include practices that use diverse rotations, biological pest control, or conservation tillage methods. Much of the debate surrounding wide scale adoption of sustainable agricultural practices, however, does not relate to their environmental impacts, rather, the debate relates to their profitability.

Profitability

Within the United States, the profitability of sustainable agriculture has been long debated [see for examples, Buttel, *et al.*, 1986; Council for Agricultural Science and Technology, 1990; Crosson and Ekey, 1988; Dobbs, 1993; Dobbs, 1994; Fox, *et al.*, 1991; General Accounting Office, 1990; Natural Research Council, 1989; and Tweeten, 1992]. A careful reading of the arguments and the evidence suggests that there are sustainable practices that do reduce negative environmental impacts from conventional practices in some, but not all, circumstances. Some, but not all, are profitable or relatively low cost[19].

A study by the Northwest Area Foundation (1994) based on six years of research in six states[20] also found that United States' farmers using sustainable practices produced a wider range of crops and livestock than did conventional farmers, but used practices consistent with those identified by the National Research Council. These farmers had more of their land in pasture, woodlands, wetlands, or other noncrop uses and used fewer purchased inputs. In general, the sustainable farmers used two

---

[18] "Sustainable agriculture" can be considered a subset of "sustainable development", that is, development that meets "the needs of the present [generation] without compromising the ability of future generations to meet their own needs" (World Commission on Economic Development, 1987). For a good discussion on concepts and issues in sustainable agriculture see *Sustainable Agriculture: Concepts Issues and Policies in OECD Countries*, OECD, 1995.

[19] It is interesting to note that, to the extent a sustainable practice significantly increases profits, it tends to become widely adopted and hence labeled conventional agriculture. Examples within the United States include conservation tillage or integrated pest management. Thus, the set of sustainable practices that have not yet become widely adopted may be heavily criticized for their profit implications. What is obscured by this labeling is the success of some of these practices - at one time labeled sustainable - in reducing negative agro-environmental services.

[20] These states were Minnesota, Iowa, North Dakota, Montana, Oregon, and Washington.

strategies: reducing chemical dependency and using ecological practices. They reduced their use of chemicals by using animal manures or soil enriching plant material to build soil quality, they applied purchased inputs on an as-needed basis, and they substituted mechanical tillage for chemical weed control. The ecological practices followed by these farmers included rotating crops to break pest cycles and enhance soil quality, integrating livestock and crop production, managing landscape by alternating strips of soil conserving and soil depleting crops, and planting trees, using windbreaks and contour farming. These practices sometimes, but not always, caused some loss of yield in some crops, but they tended to have a positive effect on the environment. While in many cases conventional farms reported higher net earnings, many sustainable farms had positive profits. Even those with reduced yields often had offsetting reduced costs. The sustainable farm that were the most profitable tended to be smaller than the other sustainable farms. Most sustainable farms also required a high degree of management skill. The study also noted that the practices required to sustainably farm varied by climate, soil type, region, and location. They also vary over time as the condition within which each farm changes.

Dobbs (1994) notes that there is a difference between analyses of the profitability and ecological implications of changing farming methods by adopting a single practice or few practices and changing a whole farming systems. In considering reviews of profitability comparisons of conventional with sustainable agriculture such as those by Fox *et al.* (1991) or Cacek and Langner (1986), he noted their inconsistent results. Profitability findings depended not only on variations in production systems and crops produced, but also on weather, soil type, and assumptions about costs and structure.

Dobbs argues, however, that if distinctions are made between practices and systems and between low input and organic agriculture, patterns of conclusions emerge from these United States' case studies. "Sustainable systems at present appear more competitive with conventional systems in predominantly small grain areas, or in transition areas between the Corn Belt and small-grain areas, than in the Corn Belt [Dobbs, 1994]. In another study, Dobbs (1993) elaborates these conclusions. He finds sustainable systems are more likely to be competitive with conventional systems in the western drier, wheat growing areas than in the higher rainfall areas such as the Corn Belt.

Any comparative study must be interpreted carefully. Profitability will be influenced by existing institutions such as farm programs, by available technology, and by demand and supply forces. For example, a change in technology that improves yields and lowers production costs of sustainable systems will change a study's conclusions as to profitability. Since research on sustainable system technologies is relatively recent, it is not unrealistic to expect such improved technologies in the near future. Indeed, a search for such technology may be induced by changes in property rights with respect to agriculture's use of the environment. Similarly, farm program reform can alter the competitive advantages between regions, as well as cropping practices and crop mix. Similarly, consumer food preferences may shift toward lower pesticide residues. Such changes alter profitability of different systems. A practice or system that is profitable in one region or institutional setting may not be profitable in another.

However, private profitability does not necessarily equate with an economically beneficial situation. Economically beneficial situations are those where the net benefits from agriculture, including all agro-environmental services, are positive. Negative agro-environmental services such as pollution can mean that a profitable farming operation is, nevertheless, not economically beneficial. Similarly, positive agro-environmental services such as visual amenities can mean an unprofitable farming operation is, nevertheless, economically beneficial. Gaps between private profitability measures and public assessments of economically beneficial situations may imply a role for public programs[21].

Uncertainty

There is considerable uncertainty as to the improvement in agro-environmental services that come from either the adoption of sustainable practices or sustainable systems, despite numerous case-specific studies[22]. The research issues in answering these questions are complex and conditions vary substantially from case to case. There are many uncertainties, and there is a need for more monitoring better modelling, and the development of environmental indicators to better assess the linkages between agricultural practices and agro-environmental services[23].

While exact measures are few and many uncertainties remain, there is nevertheless knowledge about those practices and systems most likely to provide environmental benefits - that is, those practices and systems most likely to move outcomes toward particular environmental goals. For example, the matching of the known properties of pesticides to site conditions could reduce off-site damages. The use of more efficient irrigation strategies could reduce groundwater and surface water depletion and reduce polluted return water flows. Using buffer zones to reduce the total volume and energy of cropland runoff could improve water quality. Preserving windbreaks could reduce wind erosion and provide wildlife habitat as well as visual appeal.

More is known about achieving these environmental goals than is applied. The reasons for this lack of application are many and complex; the reasons can span from lack of information or management skills to countervailing economic signals coming from both markets and agricultural policies.

3.   The role of agricultural policy

Agricultural policies have significantly influenced the magnitude and nature of the services provided by agriculture. In developed countries, agricultural policies have attempted to improve farm income by stimulating demand expansion (such as through export programs); controlling supply (such as through short and long term acreage restrictions); through price and income supports (such as direct payments); through risk reducing programs (such as crop insurance), or through subsidised inputs (such as for irrigation water).

Agricultural Producer Subsidy Equivalents[24] calculated by the OECD and the United States Department of Agriculture indicate that the developed countries have been quite supportive of their

---

[21]   See Batie (1994) for a more detailed discussion of the nature of these gaps and resultant implications for voluntary or regulatory agro-environmental policies.

[22]   For good reviews of these case studies, see both the Office of Technology Assessment (1985) and the National Research Council (1993).

[23]   The development of agro-environmental indicators is still in an early stage. See *Developing Indicators for Environmental Sustainability: The Nuts and Bolts* [Batie, 1995] for a review of various attempts to develop such indicators in Canada, Australia, the U.S. and the OECD.

[24]   The extent of government transfer payments to agriculture is measured by Producer Subsidy Equivalents or PSEs. They can be thought of as the amount of compensation that would be required to maintain farmers'

farm sectors over time. Government involvement in agriculture in non-European OECD countries has been relatively large, with Japan providing the most transfers. However, such direct governmental support is diminishing. Among Australia, Canada, Japan, Mexico, New Zealand and the U.S., only Australia's PSE did not show a decline in the past decade. (Australia's support for agriculture has been relatively low and stable over this period.)

But, tracing the impact of these programs on environmental attributes is exceptionally difficult. There is limited information as to programs' effects on farmers' choices of crop and animal enterprise mix, production practices, or conservation practices. Because agricultural programs' predominant objectives have been to raise and stabilize farm income, environmental (including landscape) impacts have not particularly relevant. Thus, environmental impacts - both positive and negative - were unintended and for most of history, largely ignored[25].

The question of the impact of federal agricultural programs on the environment is essentially one of the quantity and quality of various environmental attributes with and without the programs. However, the complexities of the agricultural economy preclude the modelling of the overall "without" situation - that is, the conditions that would have existed with respect to environmental quality had the programs never existed. Therefore, this particular question has been approached with case studies, anecdotal evidence or partial analyses. The latter are frequently conceptual and theoretically based as opposed to empirical.

The conceptual argument is that these programs, for much of their history, have altered incentives for farmers, and they result in distortions that can harm the environment. The incentives given producers by most federal farm programs are to use more inputs, to restrict rotations, and to increase yields of program crops or to stock subsidised livestock [Erdman and Runge, 1990; Kuch and Reichelderfer, 1992; Parris and Melanie, 1993; Reynolds, Moore, Arthur-Worsop and Storey, 1993 and Young and Painter, 1990]. The impacts are exacerbated by other incentives that alter income and price risk, such as subsidised insurance or transportation, or those which encourage the conversion of wetlands to agricultural uses [Reichelderfer, 1990; Phipps, Rossmiller and Meyers, 1990]. Also, some policies directed at (highly polluting) crops such as cotton or sugar have resulted in crops being grown in regions that would not have grown them to the same extent had a free market existed. The conclusion usually drawn is that high levels of support in agriculture tend to intensify the adverse effects of agricultural production on environmental quality. In addition, many conclude that farm programs have accelerated the consolidation and industrialisation of agricultural enterprises, reducing the number of smaller, more traditional and frequently more visually appealing landscapes [Heffernan, 1984; Browne, *et al.*, 1992].

Thus, "the essential explanation for these environmental impacts is that markets neither penalise farmers for them, nor offer rewards for avoiding or reducing them. Government attempts to do so, such as the United States Conservation Reserve Program, have had only limited effect. Overriding these efforts are government price support payments which have encouraged a commodity mix narrower than would be the case if payments were not restricted to certain crops, and have promoted high levels of water, fertilizer, and chemical use [Runge, 1994].

---

incomes in certain agricultural sectors, if government policies affecting agriculture(both agricultural and trade policies) did not exist [Office of Technology Assessment, 1995].

[25] Part of this section draws directly from Batie, 1984.

The case studies and theoretical arguments are convincing in the aggregate - that is, that farm programs have had negative environmental impacts, even if unintended. Nevertheless, there are caveats to this conclusion, particularly with respect to drawing implications as to the impact on the environment of agricultural policy in specific regions. The influence of agricultural policy on the environment depends on farmer participation rates, land quality attributes, and farming practices.

Participation

In different OECD countries, different commodities are included in farm programs; not all commodities benefit or do not benefit to the same extent. For example, average PSEs exceed 50 per cent of gross receipts for Canadian dairy (69 per cent), for U.S. sugar (59 per cent), and for Mexican corn and barley (57 per cent) [Nelson, Simone, Valdes, 1995]. If livestock, fruit, or vegetables are excluded from program coverage, as some are in the United States, then a major source of pollution is not directly affected by the existence of farm programs. Furthermore, not all eligible farmers of program commodities participate. Westcott (1993) notes that the portion of United States agricultural production covered by government income support payments declined the last half decade and more planting decisions are being made on the basis of market signals since the 1985 Farm Bill.

In contrast, New Zealand's past program support was directed mainly toward pastoral farmers via input subsidies for fertilizer and irrigation water, land clearance subsidies and price support for wool, beef, sheep meat and dairy products in farm programs. Since much of New Zealand's agriculture is pastoral, the past farm programs influenced the majority of practices.

Land quality

Land attributes are also crucial in assessing programs on agro-environmental services. As noted before, one argument with respect to the negative impact of agricultural policies on environmental services is that they provide incentives to use more chemical and other agricultural inputs more intensively. With a commodity policy therefore: (i) output is overvalued relative to inputs; and (ii) other agro-services such as pollution are undervalued [Antle and Just, 1991]. One conclusion is that more pesticides will be used and pollution will be made worse. A similar argument applies to water depletion, irrigation and energy use, and return water flows. Furthermore, the values associated with agro-amenities will be neglected. However, this conclusion depends on the "held assumption" that increases in chemical use automatically mean more pollution or more irrigation. Antle and Just (1991) and Just, Lichenberg, and Zilberman (1991) show that the validity of this "held assumption" depends on the environmental attributes of the land, such as erodibility.

There are also interactions with other policies - such as federal, state, or provincial environmental legislation - that can negate the "held assumption". Just and Antle (1990) note that, when one considers both agricultural and environmental policies, the policies together can have either a positive or negative effect on nonpoint pollution. "Both agricultural production and environmental impacts depend on highly location-specific environmental conditions. Reality is far too complex to allow generalisations about the environmental impacts of agricultural policies".

Farming practices

Similar caveats apply to the argument that agricultural subsidies promote the expansion (at the extensive margin) of high chemical using, highly erosive, commodity program crops such as maize [Reichelderfer, 1990]. As Thurman (1995) notes, the validity of this conclusion depends on knowing which land was converted to maize production as a result of program incentives. If those lands were already highly polluting, then conversion to maize - particularly if planted with conservation tillage and other conserving practices - may not result in increased pollution. Furthermore, real policies have historically included supply control mechanisms such as land retirement requirements. If such retirements encourage more chemical use on the farmed lands, there may not be a reduction in input use[26], but if that is not the case, then there may be an offsetting land retirement to counteract the land expansion incentive [Collins and Vertress, 1988].

Policy approaches also differ by country, by crop, and by livestock sector. Policy reform therefore will have differential impacts on agro-environmental services. For example, policies related to marketing assistance and infrastructure tend to be more important in Canada than in either Mexico or the United States [Nelson, Simone, Valdes, 1995]. Input subsidies for fertilizer, irrigation, and fuel tend to be more important in Mexico [Nelson, Simone, Valdes, 1995].

4. Policy reform

The above caveats foreshadow the difficulties in predicting the environmental effects of policy reform and trade liberalisation. As agricultural subsidies diminish, there are three basic types of impacts on agricultural production: (i) output substitution impacts; (ii) output price impacts; and (iii) input substitution impacts [Carr, *et al.*, 1988]. If there were no longer any subsidies capitalised into agricultural land values, and if there were no countervailing forces, there would be a shift in relative net returns between program and nonprogram crops. One would expect to see an output substitution impact of shifting production to nonprogram crops. If these crops are less environmentally damaging because of fewer chemical residuals, and if they are on land that is of low erosive capacity, then there could be environmental quality improvements such as water quality.

The output price effect that would be expected with policy reform is for returns for program crops to fall, however the shift to nonprogram crops would depress these nonprogram crop prices as well. Thus, one would also expect a decline in land values and producer net worth [Carr, *et al.* 1994][27]. The input substitution effect would result from the reductions in prices and land values, and would reduce the (marginal product) value to be obtained by the use of inputs. Hence use of inputs such as chemicals and fertilizer would fall, and land would be used less intensively [Carr, *et al.*, 1994].

Also, the expectation for policy reform is for an accelerated restructuring of agriculture via the elimination of the less competitive farmsteads. Visual landscapes may be altered in some areas. In areas with marginal lands which have agriculture as their only viable industry, policy reform may lead to depopulation [Weiss, 1992], if land values are less than the capitalised value of program subsidies. If the landscape amenities of the region, such as the agro-pastoral landscapes, require land

---

[26] Gardner (1991) correctly notes that the final outcome will depend on whether the Allen elasticity of substitution between land and chemicals is greater than the elasticity of demand for farm output.

[27] This conclusion assumes policy reform is unilateral rather than multilateral.

management, and if reform reduces land management, then these agro-pastoral landscapes values may be reduced - with both ecological and amenity impacts.

Actual experiences

As Carr, *et al.* (1994) note, the expected net effect on environmental quality of policy reform would probably be positive. However, if certain conditions are not meet, or if other factors enter in, then environmental improvements may not occur. The answer to the question "Does policy reform improve environmental quality?" appears to be: "It depends". Whether environmental quality improves with policy reform depends, for example, on the extent to which farmer incentives are really changed and farmer's expectations as to whether such changes are permanent. In the United States in 1985, farm programs were changed in that the yields on which payments were made were "frozen" to the 1981-85 levels. Several analysts [e.g. Reichelderfer, 1990; Gardner, 1994; and Carlson, Garduilo and Lin, 1994] argue that the subsequent decline in the aggregate use and per acre use of chemicals was a result of this 1985 policy change. Erdman and Runge (1990) disagree and argue that because farmers still register their yields for possible future calculations, the discouragement of chemical use is modest at best. Similarly, the 1996 United States agricultural policy reform still couples payments to the ownership of land. The value of decoupled, direct payments should still be capitalised into land values over time, reducing the rate of any depopulation.

Policy reform impacts on environmental quality can be swamped by non-policy related events. For example, the United States 1996 farm program reform is taking place in a period of record high prices for maize, wheat, and soybeans. Because of large increases in export demand, small grain reserves, and reduced idled land, program crop plantings are increasing, not decreasing. Early estimates are that the United States will have over 10 million acres more acres of maize planted in 1996 than in 1995 [Womack, 1996]. The impact of such increased planting will depend on the quality of the land coming into production, and the nature of production practices used on the crops that the maize is replacing. Furthermore, the high price of maize may cause producers to reduce their use of conserving practices such as filter strips, or they may encourage the removal of fencerows.

One response to the high grain prices and low cattle prices, however, is a large slaughter of the existing beef cattle herds in the United States. Animal waste is a serious pollutant in many regions of the United States. The decline in cattle numbers may have an offsetting environmental improvement to the environmental degradation that may occur from higher maize plantings. This improvement however is temporary, since livestock revenues are expected to rebound within two years.

Mexico provides another example where the impact on the agro-environmental services from policy reforms and trade liberalisation are not obvious - in part because of a Peso devaluation and exceptionally dry conditions in over 70 per cent of the Mexican Territory. Fertilizer price increases of up to 50 per cent have occurred; price increases coupled with poor sorghum and maize crops have resulted in a drawdown of cattle numbers [United States Department of Agriculture, 1995]. In short, macroeconomic events have overwhelmed agricultural and trade policy effects.

New Zealand provides one of the longest experiences with policy reform. In the mid-1980s, New Zealand unilaterally deregulated key sectors of its economy and abolished many sector-specific agricultural programs [Bollard, 1992]. The economy wide reforms ultimately raised interest rates and appreciated the New Zealand dollar, lowering domestic agricultural prices, but raising loan finance costs and disadvantaging agriculture against both its overseas competitors and other sectors in

New Zealand [Reynolds, Moore, Arthur-Worsop, and Storey, 1993]. Again, macroeconomic effects tended to dominate the specific microeconomic changes expected from policy reform. Initially, land prices dropped 60 per cent, and fertilizer use declined by up to 50 per cent [Spinelli, 1994]. However, by 1995, farmland prices had recovered to around 86 per cent of the 1982 value in real terms [Shepherd, 1996]. Fertilizer levels also increased - returning close to pre-reform levels [Shepherd, 1996]. There are fewer sheep and more cattle, deer and goats being produced [Sandrey, 1991].

It is difficult to generalise about the overall environmental effects of New Zealand policy reform. Reynolds *et al.*, (1993) examined the impacts on the New Zealand environment of reduced agricultural subsidies. They found that fertilizer use has substantially declined[28] as has new pasture development, and it "can be assumed that the intensity of farming in the harder hill and high country pastoral farms has decreased. However, certain farms in poor financial conditions are reported to be maintaining stock numbers but without fertilizer inputs. This degrades soil structure and fertility". They note that farmers have not had adequate income to invest in either soil conservation or the conversation of marginal pasture into forestry. Spinelli (1994) cautions: "If financial resources are not available to return land to prior habitat, increased soil degradation and pollution could occur". This possible increase in soil degradation, however, may be offset by the decline in pasture from 14.1 million hectares in 1983 to 13.5 million hectares in 1995, most of which is accounted for by an increase in commercial afforestations. Additional land is still classified as pastoral, but is in fact slowly reverting to woody vegetation and will ultimately be re-classified [MAF, 1996].

Pastoral agriculture has been the largest user of pesticides in New Zealand, and most pesticide use has declined following policy reform; however there has been an increase of the use of pesticides in the growing horticultural industry [Reynolds, Moore, Arthur-Worsop, Storey, 1993]. Pesticide sales remain near to 1984 levels [Shepherd, 1996]. While recognising earlier agricultural policies had a negative impact on the environment, Reynolds *et al.*, (1993) nonetheless conclude[29]:

> "There is little hard evidence to suggest that changes in agricultural output levels and mixture have been of direct benefit to the environment in the short-term. The major trend has been a reduction in grazing pressure, particularly in hill country pastures. The environmental benefits from this change could however be offset by more intensive dairy farming which may increase effluent discharges into waterways. Likewise the increase in intensive horticultural farms increases the possibility of pesticide contamination. The reduction in forest planting cannot be viewed as environmentally positive given the area of land that could usefully be put back into conservation forestry. However, the increased diversification of farms will reduce vulnerability of farms to income variation [Reynolds, Moore, Arthur-Worsop, Storey, 1993]".

Other concerns

In many countries, there is also the concern as to what might be the land use if, following policy reform, cropland or pasture converts out of agricultural uses. One of the more publicised situations of

---

[28] However, fertilizer use has slowly increased from its low in 1988 [MAF, 1996].
[29] New Zealand policy makers recognize that agricultural policy reform is not sufficient to achieve environmental goals and have begun to implement environmental law reform as well.

agriculture's relationship to the environment is found in the United States. Florida's Everglades, a unique wetlands system, is heavily polluted by the Florida cane sugar industry which comprises one-half of the domestic cane acreage [Thurman, 1995]. If the cane acreage in Florida should substantially decline in the future, either due to the demise of the sugar program or because of environmental regulations, the alternative use of the land would probably be residences. It is doubtful that the environment would be improved if such an alternative were to occur.

There are similar concerns in Japan. Should rice acreage significantly decline, and if the land is used for residential and commercial purposes, then landscape and environmental values may also decline. Of course, land use controls such as zoning could prevent such outcomes; but a decline in the value of agricultural uses can increase political pressure to allow other economic development regardless of zoning rules. Another concern involves changes in production technology. For example, expected future wide scale adoption of herbicide resistant crops such as "Roundup-Ready soybeans"[30] complicates predictions. There is concern, for example, that new herbicide resistant crops may increase the use of herbicides, may limit the use of filter strips and grass waterways due to herbicide drift, and may exacerbate herbicide carry-over from one season to the next, thus reducing the use of rotations.

Studies

There are studies that examine the impacts of United States policy reform on various agro-environmental services. These studies are cited by Ervin (1996) and he concludes that estimated reductions in U.S. environmental stresses are uneven and modest. Ervin's own study provides a good example of what is meant by uneven and modest outcomes. Ervin *et al.*, (1991) reported on the result obtained from a simulation model and concluded that, in the United States, complete flexibility as to planting would not reduce chemical use significantly. Nitrogen fertilizers would decline only about 4 per cent and pesticide use would fall more than 2.5 per cent. However, the results varied widely by region. They found that water erosion would decline in Appalachia and the Corn Belt but would increase in the Delta area of the South due to an increase in harvested area and a shift to more erosive crops.

Ervin (1996) concluded: "The removal or decoupling of subsidies lessens production pressures that exacerbate environmental damages and limit amenity values. Thus, agricultural policy reform enhances the potential for environmental improvement, but does not assure it. This conclusion is reinforced by Doering (1991): "Federal policies towards agriculture do not appear to play a major role as an incentive or a disincentive for less intensive and more environmentally benign agricultural practices and cropping systems". Parris and Melanie (1993) expand this conclusion to Japan: "While agricultural policy reform could redirect the extent and severity of the adverse impacts of Japanese agriculture on the environment, it is unlikely to provide a complete solution". Thus, agricultural program reform may facilitate but does not assure wide scale adoption of sustainable practices.

The impact of policy reform on landscape amenities and rural ownership and land use patterns is difficult to ascertain. Smith, *et al.* (1996) examined the expected farm-level economic impacts of the United States farm program changes on 71 representative crop and livestock operations located in major United States production regions and found that downward trends in net cash farm incomes were of concern for at least 24 of the farms - including some feedgrain, dairy, and wheat farms. Rice

---

[30] "Round-up Ready" features allow the use of the herbicide Round-up after plant emergence.

farms faced the greatest economic pressure over the next seven years. Livestock are expected to do reasonably well. The stressed farms may undergo changes - to different ownership, crop-mix, practices, or financing. However in many countries including the United States, many farmers depend on off-farm incomes which can cushion adverse farm economic conditions. The ultimate aggregate result in the structure of agriculture, in landscape amenities, in enterprise diversity, or even regional location of crops is difficult to predict with accuracy.

These studies and conclusion suggest that, in many cases, the enhancement of positive agro-environmental services (or the mitigation of disservices) will require more than agricultural policy reform or trade liberalisation. While policy reform has the potential to improve environmental quality, it is unlikely to achieve desired levels of improvements (i.e. mitigating damages) or enhancing the positive agro-environmental services without other environmental policies. That is, "free" markets should not be equated with "laissez faire" markets if environmental goals are to be obtained[31,32]. At the same time the achievement of environmental goals without policy reform may be frustrated if support programs entrench environmentally damaging methods of farming.

5.   The role of market forces

One phenomenon that is beginning to affect agriculture is the business-led, consumer-driven pursuit of environmental goals [Porter and van der Linde, 1995]. This trend will more likely affect habitat and ecological services than it will affect amenity services such as landscape. Many firms, particularly firms serving world markets, are encouraging or requiring the adoption of more sustainable practices by their suppliers. This change is motivated in some cases by lower production costs that can result from increased efficiency. In other cases, this pursuit appears to be driven by fear of liability or regulation. But, more often, it appears to be because the firms' managers see profitable market niches.

In the United States, for example, Gerber - a U.S. based firm that produces baby foods and whose motto is "Because it is for babies, it has to be better" - requires its contracted growers to use practices that will result in zero detection of pesticides in the final baby food product. Sainsbury's, a United Kingdom based food retailer, has started a policy of encouraging Integrated Crop Management Systems, and it provides a Code of Practices to all its suppliers. These firms, and many other like them, see a logic that links the environment, resource productivity, innovation and competitiveness. They are entering what some refer to as the third generation of environmental management where

---

[31] There are many such policies that either enhance the positive role of agriculture or which provide pollution prevention (e.g. conservation easements; management agreements; taxes, income support/green payments; regulation and land use controls), [OECD, 1989; OECD, 1993; Young, 1990]. Policies to protect the rural economic landscape include non-agricultural business development, adding value to traditional agricultural products, encouraging regional cooperation and changing rural institutions [Drabenstott and Smith, 1995]. See Creason and Runge (1989) for a general discussion of the issues of achieving both competitive and environmental goals.

[32] The role of government in pursuing agricultural or environmental goals is often the subject of contentious debate. Not infrequently, parties to the debate will - perhaps without intent - compare an "actual" status quo situation (e.g. supported agriculture and attendant agro-environmental impacts) with a theoretical ideal (e.g. unsupported agriculture with perfect markets, perfect information, and no externalities). "Actual" situations will almost always be found wanting if compared to theoretical "ideals". Debates are better informed where they compare either two theoretical ideals (e.g. "perfect" government programs with "perfect" markets) or two actual situations (e.g. "actual" government program outcomes, with "actual" market outcomes).

protecting the environment is seen as an essential element of a company's strategic approach. Such environmental management is seen as giving a firm a competitive advantage and as enhancing the firm's image. Pollution is seen as a flaw in either product design or production processes which calls for a system redesign.

The ISO 14000 certification process could conceivably codify these efforts through eco-labelling so that a purchaser will know if the product has been produced under ISO 14000 certification standards. The ISO 14000 program includes environmental management systems, environmental auditing standards, environmental performance evaluations, life cycle assessments, international labelling standards, and a registration process. The program addresses a company's entire suite of activities, from product design, planning or training and operations. It relies heavily on life cycle analysis, including cradle-to-grave, waste stream, mass balance assessments and product recycling. Some customers may require ISO certification for domestic and imported products, thus accelerating the search for enhanced agro-environmental services. A by-product of these forces is an incentive to discover complementary technologies - a form of induced innovation.

Where there are adequate market incentives, firms will adopt total environmental quality management techniques that include sustainable practices independent of or even in spite of regulation. In other cases, responding to existing or foreseeable environmental legislation will provide the needed incentive [Porter and van der Lende, 1995]. If such legislation provides flexibility as well as performance standards, firms will innovate and discover their own ways to enhance the positive agro-environmental services. These so-called "private regulations" are increasingly important; and policies that are directed toward enhancing agro-environmental service should be informed by the factors that lead to business-led and consumer-driven efforts in order to capitalise on their existence.

6.   Summary and conclusions

Agriculture provides both positive and negative agro-environmental services. There is knowledge - albeit imperfect - about the nature, magnitude, and value of agro-environmental services, about locational and enterprise differences with respect to these services, and about the feasibility of sustainable practices and systems. Less is known about the profitability from the farmer or firm standpoint of making changes. However, public policies can alter these economics by changing the property rights to the environment and by altering the institutional setting.

While the performance of agriculture is influenced by agricultural policies, and while policy reform should enhance the potential for environmental improvement, the reform of agricultural policy and trade liberalisation will not necessarily by themselves result in the attainment of positive agro-environmental services. While there are also business-led, consumer-driven market forces leading to self regulation in agriculture, many of these forces require a threat of environmental legislation or liability before these efforts appear profitable. Thus, there is a role for environmental policy to better pursue environmental goals. After policy reform, there remains a gap between farmers' economic interests in commodity values and society's interest in environmental values [Bradshaw, 1995]. The challenge to policy makers, then, is to what extent and how to close this gap.

7.   Bibliography

ANTLE, John M. and Richard E. Just (1991), "Effects of Commodity Program Structure on Resource Use and the Environment". pp. 97-128. In R.E. Just and N. Bockstael (eds.). *Commodity and Resource Policies in Agricultural Systems*. New York: Springer-Verlag.

BATIE, Sandra S. ed. (1995), *Developing Indicators for Environmental Sustainability: The Nuts and Bolts*. Proceedings of the Resource Policy Consortium Symposium, Washington, D.C. (June 12-13), Special Report (SR)89, East Lansing, Michigan State University.

BATIE, Sandra S. (1994), "Designing a Successful Voluntary Green Support Program: What Do We Know?" pp. 74-94. In Sarah Lynch (ed.) *Designing Green Support Programs*. Policy Studies Program Report #4. Washington, D.C.: Henry A. Wallace Institute.

BATIE, Sandra S. (1984), *Agricultural Policy and Soil Implications for the 1985 Farm Bill*. Occasional Paper. Washington, D.C.: American Enterprise Institute for Public Policy Research.

BATIE, Sandra S. and Daniel B. Taylor. (1991), "Assessing the Character of Agricultural Production Systems". *American Journal of Alternative Agriculture* 6(4): 184-187.

BOLLARD, Alan. (1992), New Zealand: Economic Reforms 1984-1991. San Francisco: International Center for Economic Growth, ICS Press.

BRADSHAW, Ben. (1995), "Implications of Reduced Subsidies for Agriculture and Agro-Ecosystem Health". Discussion Paper #2. Guelph, Ontario: Faculty of Environmental Sciences, University of Guelph.

BROMLEY, Daniel W. (1996), "The Environmental Implications of Agriculture". Draft paper prepared for OECD Seminar on Environmental Benefits from Sustainable Agriculture: Issues and Policies, June 18. Paris, France: OECD.

BROWNE, William P., Jerry R. Skees, Louis E. Swanson, Paul B. Thompson and Laurian J. Unnevehr. (1992), *Sacred Cows and Hot Potatoes: Agrarian Myths in Agricultural Policy*. Boulder, CO: Westview Press.

BUTTEL, Frederick. H, *et al.* (1986), "Reduced-input Agricultural Systems: Rationale and Prospects". *American Journal of Alternative Agriculture* 1(2): 58-64.

CACEK, T. and L.L. Langner (1986), "The Economic Implications of Organic Farming". *American Journal of Alternative Agriculture* 1(1):25-29.

CARLSON, Gerald A., Carlos Gargulo and Biing H. Lin. (1994), April, "The Feed Grain Program Does Not Cause Lower Crop Rotation or Higher Pesticide Use". Unpublished Manuscript. Raleigh, NC: North Carolina State University.

CARR, Barry, Klaus Frohberg, Hartley Furtan, S.R. Johnson, William H. Meyers, Tim Phipps, and G.E. Rossmiller (1988), "A North American Perspective on Decoupling", pp. 113-140. In William M. Miner and Dale E. Hathaway (eds.) *World Agricultural Trade: Building a Consensus*. Halifax, Nova Scotia: Institute for International Economics.

COLLINS, Keith and James Vertrees (1988), "Decoupling and U.S. Farm Policy Reform". *Canadian Journal of Agricultural Economics* 36(4) part 1:733-745.

Council for Agricultural Science and Technology. (1990) July, *Alternative Agriculture: Scientists' Review*. Special Publication No. 16. Ames, Iowa.

CREASON, Jared R. and C. Ford Runge (1989), *Agricultural Competitiveness and Environmental Quality: What Mix of Policies Will Accomplish Both Goals?* St. Paul, MN: University of Minnesota, Center for International Food and Agricultural Policy.

CROSSON, Pierre and Janet Ekey (1988) November, *Alternative Agriculture: A Review and Assessment of the Literature*. Discussion Paper ENR88-01. Washington, D.C.: Resource for the Future.

DOBBS, Thomas L. (1993), Implications of Sustainable Farming Systems in the Northern Great Plains for Farm Profitability and Size. Economics Department Staff Paper 93-3. South Dakota: Brookings.

DOBBS, Thomas L. (1994), "Profitability Comparisons: Are Emerging Results Conflicting, Or Are They Beginning to From Patterns?" Paper presented at the American Agricultural Economics Association Annual Meeting, San Diego, California. August 8.

DOERING, Otto (1991), "Federal Policies as Incentives or Disincentives to Ecologically Sustainable Agricultural Systems" A Presentation to the Environmental Protection Agency Sustainable Agriculture Workshop. Washington D.C. July 22.

DRABENSTOTT, Mark and Tim R. Smith. (1995), pp. 180-196. In Emery N. Castle (ed.). *The Changing American Countryside: Rural People and Places*. Manhattan, KS: University Press of Kansas.

EDGELL, Michael C. and David E. Nowell (1989), "The New Environmental Paradigm Scale: Wildlife and Environmental Beliefs in British Columbia". *Society and Natural Resources* 2(4): 285-296.

ERDMAN, L. and C.F. Runge (1990), "American Agricultural Policy" and the 1990 Farm Bill". Review of Marketing and Agricultural Economics 58(2,3):109-126.

ERENSTEIN, Olaf. (1994), "Are Productivity Enhancing Resource Conserving Technologies a Viable "Win-Win" Approach in the Tropics? The Case of Conservation Tillage in Mexico". pp. 313-322. In William Lockeretz (ed.). *Environmental Enhancement Through Agriculture*. Proceedings of a Conference held at Tufts University in Boston, Massachusetts. November 15-17.

ERVIN, David E. (1996), "Agriculture, Trade and the Environment: Anticipating the Policy Challenges". Draft document submitted to the Joint Working Party of the Committee for Agriculture and the Environment Policy Committee, Organization for Economic Co-operation and Development. Paris, France. June 19-21.

ERVIN, David, Kenneth Algozin, Marc Carey, Otto Doering, Stephen Frerichs, Ralph Heimlich, Jim Hrubovcak, Kazim Konyar, Ian McCormick, Tim Osborn, Marc Ribaudo and Robbin Shoemaker (1991), *Conservation and Environmental Issues in Agriculture*. Staff Report #9134. Washington D.C.: Economic Research Service, United States Department of Agriculture.

FITZPATRICK, Kerry J. (1995), "Birds of Prey and Their Use of Agricultural Fields". pp. 103-112. In William Lockeretz (ed). *Environmental Enhancement Through Agriculture*. Proceedings of a Conference Held at Tufts University in Boston, Massachusetts, November 15-17.

FOX, Glen, Alfons Weersink, Ghulam Sarwar, Scott Duff, and Bill Dean (1991), "Comparative Economics of Alternative Agricultural Production Systems: A Review". *Northeastern Journal of Agricultural and Resource Economics*. 20(1):124-42.

GARDNER, Bruce L. (1991), "Redistribution of Income Through Commodity and Resource Policy". pp. 129-142. In R.E. Just and N. Bockstael (eds.). *Commodity and Resource Policies in Agricultural Systems*. New York: Springer-Verlag.

GAO [General Accounting Office] (1990) February, *Alternative Agriculture: Federal Incentives and Farmers' Opinions*. GAO/PEMD-90-12, Program Evaluation and Methodology Division. Washington, D.C. (February).

HAYS, Samuel P. (1958), *Conservation and the Gospel of Efficiency*. United Kingdom: Cambridge Press.

HEFFERNAN, William D. (1984) May, "Examining the Consequence of Recent Agricultural Policy on Farm Families in Rural Communities". pp. 1-22. In Sandra S. Batie and J. Paxton Marshall (eds.). *Restructuring Policy for Agriculture: Some Alternatives*. College of Agriculture and

Life Sciences Information Series 84-2. Blacksburg, VA: Virginia Polytechnic Institute and State University.

ITTELSON, William H., Harold M. Proshansky, Leanne G. Rivlin, Gary H. Winkel and Dempsey (1974), *An Introduction to Environmental Psychology*. New York: Holt, Rinehart and Winston, Inc.

JUST, Richard E. and John M. Antle (1990), "Interactions Between Agricultural and Environmental Policies: A Conceptual Framework. *American Economics Association Papers and Proceedings*. 80(20):197-202.

JUST, Richard E., Erik Lichtenberg and David Zilberman (1991), "Joint Management of Buffer Stocks for Water and Commodities". pp. 173-195. In R.E. Just and N. Bockstael (eds.). *Commodity and Resource Policies in Agricultural Systems*. New York: Springer-Verlag.

KUCH, Peter and Katherine Reichelderfer (1992), "The Environmental Implications of Agricultural Support Programs: A United States Perspective. pp. 215-231. In T. Becker, R. Gray, and A. Schmitz (eds.). *Improving Agricultural Trade Performance Under the GATT*. Wissenschaftsverlag Vauk Kiel KG. Germany: Kiel.

MAF Policy, (1996), "New Zealand: The Environmental Effects of Removing Agricultural Subsidies". Paper prepared for OECD Seminar on Environmental Benefits from Sustainable Agriculture: Issues and Policies. September 10-13. Helsinki, Finland.

MURPHY, William, Joshua Silman, Lisa McCrory, Sarah Flack, Jon Winsten, David Hoke, Abdon Schmitt, and Brian Pillsbury (1995), "Environmental, Economic, and Social Benefits of Feeding Livestock on Well-Managed Pasture". pp. 125-134. In William Lockeretz (ed). *Environmental Enhancement Through Agriculture*. Proceedings of a Conference Held at Tufts University in Boston, Massachusetts, November 15-17.

National Research Council (1989), *Alternative Agriculture*. Washington, D.C.: National Academy Press, Board on Agriculture.

National Research Council (1993), *Soil and Water Quality: An Agenda for Agriculture*. Washington, D.C.: National Academy Press.

NELSON, Frederick J., Mark V. Simone and Constanza M. Valdes (1995) September, *Comparison of Agricultural Support in Canada, Mexico, and the United States*. Washington D.C.: Economic Research Service, United States Department of Agriculture.

Office of Technology Assessment (1995), Agriculture, Trade, and the Environment: Achieving Complementary Policies. Washington. D.C.: Congress of the United States.

Organisation for Cooperation and Economic Development [OECD] (1996), "Voluntary, Co-operative Approaches to Sustainable Agriculture". Draft paper submitted to the Joint Working Party of the Committee for Agriculture and the Environment Policy Committee. Paris, France: OECD.

Organisation for Economic Cooperation and Development [OECD] (1995), *Sustainable Agriculture: Concepts, Issues and Policies in OECD Countries*. Paris, France: OECD.

Organisation for Economic Cooperation and Development [OECD] (1994), *The Contribution of Amenities to Rural Development*. Paris, France: OECD.

Organisation for Economic Cooperation and Development [OECD] (1993), *Agricultural and Environmental Policy Integration: Recent Progress and New Directions*. Paris, France: OECD.

Organisation for Economic Cooperation and Development [OECD] (1989), *Agricultural and Environmental Policies: Opportunities for Integration*. Paris, France: OECD.

PARRIS, Kevin and Jane Melanie. (1993) September, "Japan's Agriculture and Environmental Policies: Time to Change". *Agriculture and Resources Quarterly* 5(3): 386-399.

PETRUCCI, Bryan T. (1995), "The Potential of Dairy Grazing to Protect Agricultural Land Uses and Environmental Quality in Rural and Urban Settings". pp. 145-150. In William Lockeretz (ed).

Environmental Enhancement Through Agriculture. Proceedings of a Conference Held at Tufts University in Boston, Massachusetts, November 15-17.

PHIPPS, Tim T., George E. Rossmiller, and William H. Meyers (1990), "Decoupling and Related Farm Program Options". pp. 101-124. In Kristen Allen (ed). *Agricultural Policies in a New Decade.* Washington D.C.: Resources for the Future.

PORTER, Michael E. and Claas van der Linde (1995), "Toward a New Conception of the Environment - Competitive Relationship". *Journal of Economic Perspectives* 9(4):97-118.

RAYNER, Steve (1994) February, "A Cultural Perspective on Structure and Implementation of Global Environmental Agreements". *Evaluation Review* 15(1): 75-102.

RAUSSER, Gordon C. and William E. Foster (1991), "The Evolution and Coordination of U.S. Commodity and Resource Policies". pp. 17-45. In R.E. Just and N. Bockstael (eds.). *Commodity and Resource Policies in Agricultural Systems.* New York: Springer-Verlag.

REICHELDERFER, Katherine (1990), "Environmental Protection and Agricultural Support: Are Tradeoffs Necessary" pp. 201-230 In Kristen Allen (ed.). *Agricultural Policies in a New Decade.* Washington D.C.: Resources for the Future.

REYNOLDS, Russ, Walter Moore, Murray Arthur-Worsop and Mark Storey (1993), "Impacts on the Environment of Reduced Agricultural Subsidies: A Case Study of New Zealand". MAF Policy Technical Paper 93/12. Wellington, New Zealand: Ministry of Agriculture and Fisheries.

RUNGE, C. Ford (1994), "The Environmental Effects of Trade in the Agricultural Sector" pp. 19-54. In *The Environmental Effects of Trade.* Organization for Economic Cooperation and Development, Paris, France.

SANDREY, Ron A. (1991), "Economic Reforms and New Zealand Agriculture". *Choices* (First Quarter).

SMITH, Edward G., James W. Richardson, Allan W. Gray, Steven L. Klose, Joe L. Outlaw, John W. Miller, Ronald D. Knutson and Robert B. Schwart, Jr. (1996) April, "Representative Farms Economic Outlook: FAPRI/AFPC April 1996 Baseline". AFPC Working Paper 96-1. College Station, Texas: Agricultural and Food Policy Center, Texas A&M University.

SOULE D. Judith and Jon K. Piper (1992), *Farming in Nature's Image: An Ecological Approach to Agriculture.* Washington D.C.: Island Press.

SPINELLI, Felix (1994) December, "Farming Without Subsidies in New Zealand: *Agricultural Outlook.* Washington, D.C.: United States Department of Agriculture.

The Northwest Area Foundation (1994) December, "A Better Row to Hoe: The Economic, Environmental, and Social Impact of Sustainable Agriculture". St. Paul, MN: The Northwest Area Foundation.

THURMAN, Walter N. (1995), *Assessing the Environmental Impact of Farm Policies.* Washington, D.C.: American Enterprise Institute Press.

TWEETEN, Luther (1992), "The Economics of an Environmentally Sound Agriculture (ESA)". *Research in Domestic and International Agribusiness Management* 10: 39-83.

United States Department of Agriculture (1995) September, *NAFTA: What's Up?* Washington, D.C.: NAFTA Economic Monitoring Taskforce, Economic Research Service.

VAIL, David (1995), "The Living Countryside: Maintaining Sweden's Agrarian Landscape" pp. 275-284. In William Lockeretz (ed.). *Environmental Enhancement Through Agriculture.* Proceedings of a Conference Held at Tufts University in Boston, Massachusetts, November 15-17.

VOGEL, David (1990), "Environmental Policy in Europe and Japan," pp. 257-278. In Norman Vig and Michael E. Kraft (eds.). *Environmental Policy in the 1990s: Toward a New Agenda.* Washington, D.C.: Congressional Quarterly Press.

WASHBURN, Steven P., Rene J. Knook, James T. Green, Jr., Gregory D. Jennings, Geoffrey A. Benson, James C. Barker and Matthew H. Poore (1995), Enhancement of Communities with Pasture-Based Dairy Production Systems. pp. 134-144. In William Lockeretz (ed.). *Environmental Enhancement Through Agriculture.* Proceedings of a Conference Held at Tufts University in Boston, Massachusetts, November 15-17.

WEISS, Christopher R. (1992), "The Effect of Price Reduction and Direct Income support Policies on Agricultural Markets in Austria". *Journal of Agricultural Economics* 43(1): 1-13.

WESTCOTT, Paul C. (1993) June, *Market-Oriented Agriculture: The Declining Role of Government Commodity Programs in Agricultural Production Decisions.* Agricultural Economic Report Number 671. Washington, D.C.: Economic Research Service, United States Department of Agriculture.

WOMACK, Abner (1996), Co-Director of the Food and Agricultural Policy and Research Institute, U of Missouri, Columbus, Missouri. Personal Communication. May 25.

World Commission on Environment and Development (1987), *Our Common Future.* The Bundtland Report. Oxford, United Kingdom: Oxford University Press.

YOICHI, T. (1992), "An Environmental Mandate for Rice Self-Sufficiency". *Japan Quarterly* January-March: 34-44.

YOUNG, Michael D. (1990) January, "Agriculture and the Environment: OECD Policy Experiences and American Opportunities". Policy Planning and Evaluation (PM-221). Washington, D.C.: United States Environmental Protection Agency.

YOUNG, Douglas L. and Kathleen M. Painter (1990), "Farm Program Impacts on Incentives for Green Manure Rotations". Unpublished Manuscript. Pullman, Washington: Washington State University.

ENVIRONMENTAL BENEFITS OF AGRICULTURE: EVALUATION METHODS TO MEASURE AND MONITOR CHANGE

by
W. George Hutchinson
Queen's University Belfast, UK

Introduction: The methods reviewed

This paper will review the two major methods for measuring and monitoring change in the environmental benefits from agriculture: physical and monetary evaluation methods. Discussion on physical evaluation methods will centre on the physical inventory approach for monitoring change in all significant attributes of a beneficial environmental service. The construction of physical indicators and their relationships within a Driving Force State Response. (DSR) Model will be considered. Discussion on monetary evaluation methods will centre on the Contingent Valuation Method, (CVM). It will be shown how in certain circumstances, this can be extended into a full Cost Benefit Analysis (CBA). Both physical and monetary evaluation methods should provide analytically sound measures of policy relevant environmental benefits from agriculture which are demonstrated to be of value to society. They should be able to monitor trends in benefits provided within a given region over time and should permit interregional and international comparisons of levels and trends in such benefits.

While some literature on the subject may give the impression that the two basic evaluation methods represent mutually exclusive schools of thought and alternative approaches to the subject on hand, this paper is opposed to such a view. It will, however, look at each method separately, examining strengths and weaknesses and areas of applications where that method is most appropriate. The paper will also demonstrate that there are strong logical interrelationships between the methods, with monetary measures clearly constructed on the foundation of fundamental physical measurements. Under policy recommendations, it is suggested that some serious problems in measuring benefits of landscape quality and biodiversity might be resolved if more attention was focused on the concepts and methods used to estimate the preservation value of such benefits.

1. Separating environmental benefits from environmental costs: The Reference Level Concept

While most subsequent discussion will be of positive externalities and beneficial environmental services provided by agriculture, it is necessary at the outset to explain how benefits and costs can be created by the management of both beneficial and detrimental environmental services. This is conceptually quite a complex matter but is explained in some detail by Hodge (1994). Before proceeding it is necessary to separate beneficial from detrimental environmental services of agriculture. The previous paper by Bromley (1996) has demonstrated that this is not as simple as it

might seem as a service which is beneficial to one person may be detrimental t
nevertheless a reasonable consensus that increased use of nutrients and pesticides
to increase detrimental services. Conversely, increases in carbon fixing by agric
in the quality of maintenance of biodiversity, wildlife habitat and landscape are re
beneficial services.

Strictly speaking whether the level of a beneficial or detrimental agricultu
considered as producing a cost or a benefit to society, depends on social and political judge
what are the responsibilities or duties associated with land ownership and hence the position
reference level of environmental quality related to that externality. Where landowners fail to achieve
the reference level of environmental quality by either over-providing a detrimental service, or
under-providing a beneficial service, they create an external cost as shown in Figure 1. Where they
exceed the reference level in provision of a beneficial service, or inflict less of a detrimental service
than permitted by the reference level, they generate an external benefit for society as shown in
Figure 1.

**Figure 1. How the reference level separates benefits from costs in management of
beneficial and detrimental agri-environmental services**

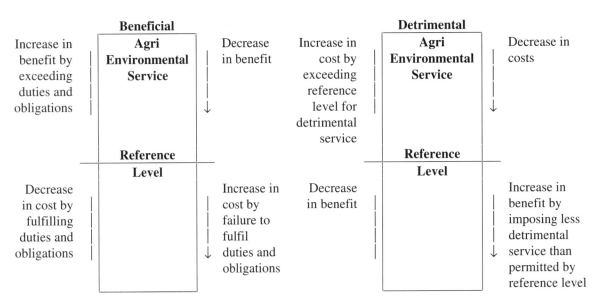

It may be observed that the reference level, for provision of beneficial environmental services, from agriculture in OECD countries, is set at significant positive levels of such services and expectations may be for a reference level of services, not distinctly less than, and in some cases exceeding those currently provided. As discussed in the previous paper by Bromley (1996), the property rights to beneficial services are a complex issue and do not necessarily belong to landowners, who may therefore not have the right to withdraw services to below the reference level, or be compensated for maintaining them at that level. The position of the expected reference level may also change over time, and in some cases there may be significant upward shifts in public expectations of acceptable reference levels, as for example, in food safety and animal welfare standards.

It follows from the above framework that only increases in beneficial agri-environmental services which exceed the reference level can be referred to as environmental benefits (Figure 1). An appropriate policy response in this situation could be to make payments to landowners in exchange for increases in benefits above the reference level. Services provided up to the reference level are not benefits, they are the duties and obligations that society and conventional land management practices place on land owners. Such services should continue to be provided without payment to landowners and, in many cases, may not be under threat of reduction as they form joint products with private goods, or arise from landowners preferred management practices. Where landowners fail to achieve reference levels enforcement including regulation and taxation, and not payments, are appropriate policy responses.

Failure to recognise this important distinction between beneficial environmental services provided below and above the reference level could result in potentially gross overestimates of benefits, irrespective of whether the evaluation is by physical or monetary measures. Such overestimates are most likely to occur when measuring absolute levels rather than the changes in benefits, which are of much greater policy interest. Caution with regard to classifying services below the reference level, as benefits would suggest, that threshold levels be set for all environmental indicators as a matter of priority.

2.   Overview and discussion of evaluation methods

Physical evaluation methods and the Driving Force State Response (DSR) Model

The physical inventory method is a data intensive method measuring levels and changes in all significant attributes of a beneficial service arising from all sources, industry, agriculture, households etc. Unless the benefit is closely related to the level of one key attribute e.g. CO_2 in the case of greenhouse gases, problems of prioritising and weighting the significance of change in several attributes become highly complex. Change in attributes may also have quite different effects over the related environmental benefit when they take place in different member states or even in different regions of a member state depending on local conditions.

Use of the method in such cases is more problematic than in cases where an attribute change has a constant universal significance, as in the case of a global issue such as CO_2 balance. Where a global inventory makes sense this method can work well in making comparisons between national and regional contributions to this inventory as in the greenhouse gases issue. In such cases, a universal physical inventory also works well in comparing contributions of all different sectors of the economy, to increasing or reducing the problem. This point is well made by Adger and Whitby (1993) who established a balance for greenhouse gases for UK land use, but conclude that unless this finding is placed in the context of the balance for the whole UK economy its relatively insignificant effect may not be fully appreciated.

The physical inventory method has recently been suggested by the European Environmental Agency (1995) as an appropriate method to monitor landscape change, by setting up a European-wide "landscape inventory" which could allow a systematic approach for assessing the dimension rate and trends of landscape changes. The task of collecting and updating such an inventory for Europe, Canada or Australia seems enormous. An effective method for monitoring landscape change over a five year period is explained in the UK Department of the Environment (1993) Countryside Survey 1990. This is based on detailed field surveys in 508 randomly selected one kilometre squares throughout Great Britain. Such sampling techniques make this method viable on a national scale. The method should work best, to monitor change over time, in landscape quality or habitat maintenance, in the same region where change takes place in the same bundle of attributes. International or interregional comparisons are less meaningful for here change is taking place in different and physically non comparable bundles of attributes. Although the physical inventory method of monitoring may look cumbersome and simplistic, the collection of fundamental physical facts form the basis on which physical and monetary evaluation methods are all ultimately based.

Physical indicators are measures based on the physical inventory method. Indicators represent components or processes of real world systems and are models of the real world which condense data on the subject of interest as shown in Figure 2. Good indicators are defined as indicators which are consistent in the representation of complex processes by a format which may aid decision makers rather than confuse them. The process of condensing data is shown in Figure 2. At the outset agriculture's contribution to the benefit is separated from contributions from other sources. Indicators of interest to scientists may still include a considerable amount of raw data on many different attributes of the benefit. Physical indicators for policy makers should be related to policy objectives, evaluation criteria, target and threshold levels and should be condensed into a few key attributes of the benefit which can be monitored over considerable periods of time to recognise clear trends. Monitoring should also take place at different levels of aggregation to investigate possible differences between regional trends or between trends for farm types or farm sizes.

Figure 2. Condensing change in the physical inventory of attributes of an environmental benefit into change in physical indicators

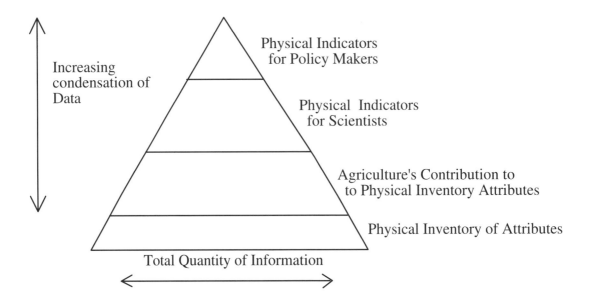

Much published work on physical indicators, both for agriculture and the general economy, has used the Pressure State Response (PSR) or "causality" Model [OECD, 1994b; and Heimlich, 1995]. This approach examines the possibility of producing indicators at three separate levels: (i) the human and economic activities that exert pressure on the environment; (ii) the environmental conditions or states that prevail as a result of that pressure; (iii) the response by society to changes in the pressure and state of the environment. A slightly modified form of this framework, known as the Driving Force-State-Response (DSR) Model used in recent OECD (1996) work on Agri-Environmental Linkages is shown in Figure 3.

It is worth commenting that, while the most valuable physical indicators of change in attributes of environmental benefits would probably relate to changes in environmental states, such as states of landscapes or ecosystems, it has proved easier to develop indicators of trends in driving forces or pressures, such as changes in chemical inputs, farm management practices and percentage of land area held in protected categories as shown in the Annex. There has been relatively little work done on indicators of responses, especially government policy responses to changes in environmental states or driving forces.

In an ideal world, modelling exercises involving simulation techniques should be able to predict the effects of change in driving forces on environmental state, and the effect of changes in policy response on environmental state and driving forces. Such economic and ecological modelling is a highly complex exercise, given the non-linearity and highly site-specific nature of many of the systems relationships. This use of physical indicators, to predict change in environmental state, is less well developed than their use to monitor change in easily quantified driving forces or pressures as shown in the Annex.

Monitoring these straightforward physical indicators of environmental pressure is most useful when tracking environmental change over time within the same region. It is however less useful for making meaningful international or even interregional comparisons. If we consider agricultural nutrient uses, as an example of an environmental driving force, we find that the effect on environmental state of change in the leading indicator, nutrient imbalance, depends quite crucially on area specific agro-ecological features, such as soil type, hydro-geological conditions, climate etc. This means that the environmental pressure represented by the indicator will only have similar effects in similar agro-ecological zones. Site specific factors also limit comparability of physical indicators measuring change in maintenance of wildlife habitat or landscape quality. Some of these problems may be solved by developing simulation techniques to predict the interaction of major site specific characteristics on indicator change. Some subjects, such as maintenance of landscape quality, appear to defy such modelling because they cannot be represented by a few key criteria. Furthermore interactions between these multiple criteria of landscape quality are not understood and subjective judgement can play an important role. Site specific factors may not however constrain comparisons of indicators of more uniform global issues such as CO_2 balances.

In cases where specific features of the site have an important interaction with the indicator, international comparisons of absolute levels of indicators are devoid of meaning unless applied to similar agro-ecological zones which are difficult to define and map and may be highly concentrated geographically. A fall back position is that changes and trends in physical indicators (such as those in the Annex) provide more meaningful international comparisons but even these are not entirely free of the influence of critical site specific interactions. It would also be useful for such comparisons to have a set of national and regional threshold and target levels for indicators to signal whether changes are taking place, above or below reference levels.

Figure 3. The Driving Force-State-Response Framework to address agri-environmental linkages and sustainable agriculture

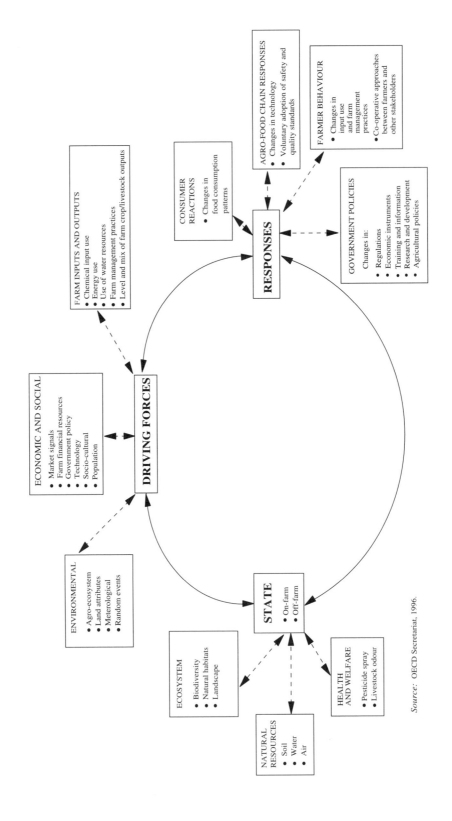

Source: OECD Secretariat, 1996.

With regard to the setting of threshold and target levels, there is some discussion in the agri-environmental literature (and a broader discussion in the sustainability literature) on the appropriate geographic and administrative level for setting thresholds and targets for physical indicators [King and Crosson, 1995; and Chilton, 1996]. The concept of the agro-ecological zone has already been discussed and this accords with the general principle that policy intervention should be at the same scale as the problem. This so-called locus of decision making has a clear social as well as scientific basis. The choice of indicators, thresholds and targets should rest with the community consuming the agri-environmental benefits, thus emphasising a principle of subsidiarity and local decision making. If, however, the benefits are the subject of regional, national or supra-national financing, then the locus of decision-making on thresholds and targets must extend to include input from the budgetary centre and those with input to funding.

There appears to be considerable interest in predicting the effect of international agricultural policy reform and trade liberalisation on indicators [OECD 1994]. General indications suggest reductions in support may reduce certain environmental costs related to use of chemical inputs and may even lead to increases in benefits resulting from less intensive land use. Suggestions include modelling to forecast the impact of change in such policy response on environmental state. This paper will consider the simpler issue of the two distinct effects on benefit indicators arising from policy reform and trade liberalisation: (i) the effect of an overall reduction in agricultural support on indicators, and (ii) the effect on indicators of a reorientation of agricultural policy to provide support directly targeted to increase Agri-Environmental Benefits.

While the effects of (i) on benefit indicators may be neutral or even ambiguous the effect of (ii) is almost certain to be positive and the more expenditure targeted under (ii) the greater the likely increase. Given the two part nature of the above process it would be desirable to separate the effects of (i) on indicators from that of (ii). This could be done by monitoring change in indicators for samples of farms which participate in and which do not participate in agri-environmental programmes under (ii). Such a process could follow the simple method used by Curtis and De Lacy (1996) to monitor differences in attitudes on environmental issues between groups of participants and non participants in the Australian Land Care Programme.

Monetary valuation approach, cost benefit analysis and cost effectiveness analysis

The review by Johansson (1994) establishes that the value of agri-environmental benefits such as maintenance of landscape quality and abundance of biodiversity and wildlife habitat is largely a non use, or preservation value and only in small part a value based on demand for recreational use. The Contingent Valuation or survey method (CVM) is therefore the most appropriate valuation technique. This method was the only method considered as currently capable of assessing non use value by a government appointed panel of leading economists in the United States [Arrow *et al.,* 1993]. The panel has further approved CVM for providing estimates of the value of environmental damage from oil spills in cases brought before the US courts.

Valuing change in specific benefits, such as landscape maintenance or habitat preservation, represent feasible but complex applications of the method and policy studies should reflect good practice as laid down by the reference manual by Mitchell and Carson (1989) and the report by Arrow *et al.* (1993). The technique is extremely flexible and is capable of providing estimates for the value or willingness to pay (WTP) to implement or avoid change in virtually all quantifiable environmental states. Results can be aggregated at different levels up to and including nation level, such as the

United States national water quality survey, [Mitchell and Carson, 1984]. Estimates can be used in international comparisons if derived using similar survey methods. Another advantage of this method is that comparisons can easily be made between the significance of changes in several physically non comparable environmental benefits without recourse to complicated weightings, as all benefits and costs are given in monetary terms.

Results from CVM studies should always be closely scrutinised as they are very dependent on the details of the questionnaire particularly the wording of the valuation scenario which explains to respondents the precise amount of the environmental change to be valued. Small changes should elicit smaller values and larger changes should elicit larger values. Willingness to pay to avoid huge postulated reductions in an environmental benefit including reductions to far below the reference level (Figure 1) or even complete removal of the service usually elicit very high estimates of value. Such wrongly specified estimates of value are of course totally non compatible with the small physical changes which are taking place in physical indicators in the real world and are therefore of little use to policy makers.

CVM is a hypothetical method and this has the advantage that a range of changes in a benefit can be valued. Rather than using purely hypothetical changes it is better for policy purposes to value changes equivalent to those taking place in physical indicators. In such cases monetary evaluation is actually based on data derived from physical evaluation. The major manuals on CVM [such as Mitchell and Carson, 1989] all suggest that monetary values for benefits be compared with the financial cost of their provision within a full cost benefit analysis (CBA). If specifically targeted agri-environment programmes are introduced offering payments or other incentives such as educational or advisory services to landowners to encourage provision or maintenance of specific benefits then benefit cost tests should be conducted to show the effectiveness of such expenditure (Table 1). Placing monetary values on benefits is expensive and time consuming and a cheaper but inferior policy appraisal method known as cost effectiveness analysis (CEA) can be used. Here an indicator of the physical amount of benefit is compared with financial cost of provision to give a cost effectiveness ratio (Table 1).

Table 1. Major economic ratios for evaluation of agri-environmental programmes

D:C	=	The Benefit Cost Ratio (for a given Programme)
B:C	=	The Cost Effectiveness Ratio (for a given Programme)
where		
D	=	monetary benefit (for a given Programme)
C	=	monetary cost (for a given Programme)
B	=	indicator of physical benefit (for a given Programme)

This method is used in the US to rank landowners' bids for providing similar services under the Conservation Reserve Programme (CRP) as discussed in Heimlich (1995). In cases where government support for agri-environmental benefits is channelled through commodity or income support programmes (which may involve cross compliance clauses) no separate agri-environment expenditures or costs can be determined. In such cases policy effectiveness cannot be assessed using CBA or CEA. For specifically targeted agri-environmental measures for which the benefit cost and cost effectiveness ratios in Table 1 are available, it would be worthwhile to compare these ratios for alternative policy approaches, such as direct payment to landowners and educational and advisory programmes to see which approach is most cost effective.

Further possible cross comparison between data from physical and monetary evaluation methods and mainstream statistics from OECD agricultural databases are given in Table 2. As with the relationships in Table 1 some of these financial relationships may be amenable to international comparison.

Table 2. Economic analysis of agri-environmental programmes

C/N	=	Expenditure per Participating Farm
C/L	=	Expenditure per hectare of land in programme
C:G	=	Expenditure to Gross Margins of Participating Farms
D/N	=	Value of Benefit per participating farm
D/L	=	Value of Benefit per hectare of land in programme
D:G	=	Value of Benefit to Gross Margins of Participating Farms
where		
N	=	number of farms in the programme
L	=	hectares of land in the programme
G	=	gross margins of participating farms

3. Policy recommendations

Effective measurement of benefits and costs by either physical or monetary methods face both theoretical and practical problems. The theoretical problems appear to be largely resolved by interpretation of Figure 1. Benefits are created only by providing levels of beneficial services above the appropriate reference level. (Costs arise similarly from detrimental services provided above their reference level.) To make theory work in practice national and regional threshold (or reference) levels need to be set.

As a theoretical prescription policy makers might consider payments, or educational and advisory services, to encourage those providing beneficial services at levels clearly above the threshold levels. Effectively this means payments and services to encourage those already providing high standards of environmental services to maintain or increase their efforts. Payments and government services are, however, inappropriate instruments to encourage under-providers below the threshold level to rise to a reasonable standard of beneficial service. Such under-providers are shown by Figure 1 to be imposing environmental costs on society by failing to fulfil their obligations and should be raised to the threshold by a variety of enforcement measures akin to implementation of the Polluter Pays Principle.

Turning to practical problems, these appear more serious than the theoretical problems and there appear to be more unresolved problems in measuring beneficial services than detrimental services. For measuring both beneficial and detrimental services the key questions are: i) How do we set the reference level? and ii) How do we measure objectively how far any region, land use type, or farming system is above/below this reference level?

Both issues are examined by investigating nutrient use (a major agri-environmental cost) and landscape quality (a major agri-environmental benefit). The approach outlined above appears to suit the nutrient use case well but transfers poorly to the landscape quality case. Looking firstly at the nutrient use case physical and monetary cost estimates are produced by following a few simple steps: (i) Acceptable regional or national reference levels are set for fresh water and drinking water quality. This is not just a scientific decision but involves social and political input. In the case of drinking water quality there are European Union norms. (ii) Physical data on the environmental impact of agricultural discharges on fresh water and drinking water quality can be compared with appropriate threshold levels, to give physical measures of environmental cost in terms of excesses above these thresholds. (iii) The contingent valuation method (CVM) can be used to put monetary values on the physical excesses, estimated in the second step, by a survey method conveying to bidders the implications of physical excesses in terms of ecosystem damage and risk to human health.

If we now attempt to impose a model similar to the costs of nutrient use on the benefits of landscape quality we encounter serious problems. It is practically impossible to set overall national or regional threshold levels for landscape quality as a large number of parameters could be involved. It is furthermore practically impossible to measure objectively how far a region, land use type or farming system, is above or below this threshold as trade-offs would have to be made between the many parameters involved and subjective judgement and personal tastes and preferences could play an important role.

As a tentative suggestion to find a way round the problem outlined it is proposed that further work on benefit indicators for landscape quality and biodiversity of the agri-environment should broaden its academic base and explore carefully the conceptual literature on preservation value, uncertainty and environmental benefits of the natural environment. The original conceptual papers on this subject are Weisbrod (1964) and Krutilla (1967). A large theoretical and applied literature has since developed mostly applied to the natural environment. This raises the question of what is the major source of agri-environmental benefit from landscape quality and biodiversity. One possible source is physical improvement in landscape quality and biodiversity resulting from increased uptake of beneficial processes such as less intensive grazing, less nutrient use or other environmentally friendly methods measured by a variety of physical driving forces. This is probably only a minor source of landscape and biodiversity benefits. The major source of benefits is almost certainly the maintenance by some agricultural sectors of high current levels of landscape and biodiversity benefits which are under risk of future decline. Such benefits have preservation value and there is a considerable willingness to pay to ensure their continued supply.

Contingent Valuation Studies have a long history of measuring the preservation value of aspects of the natural environment. In these studies the emphasis is not on the threshold level and estimates of where the current level is relative to the threshold (Figure 1) but the status quo level (current level) and the risks of decline in this level. For conceptual discussion of these issues in an agri-environment context see Willis *et al.* (1993) and Hutchinson *et al.* (1995).

Simple examples related to landscape and biodiversity benefits can illustrate this point. Specific regions or landuse types such as low intensity farming systems can be identified where the status quo level of landscape quality or biodiversity significantly exceeds average levels in the member state. (Sometimes this identification can be based on independent scientific evidence such as prior designation as an Area of Outstanding Natural Beauty (AONB) or area of scientific interest.) Expert assessment should be made of the risks to landscape quality or biodiversity in these cases and any decline which could take place as a result of agricultural change over a given time scale. Provided that the post change state is still considered to be above the threshold level, then the physical difference between benefit attributes provided at the status quo level and the post change level is a physical measure of benefits currently provided, which are uncertain to continue. The contingent valuation method can be used to estimate monetary measures of willingness to pay to preserve such "at risk" agri-environmental benefits. If as argued agricultural landscape quality and biodiversity are accepted as having mostly preservation or maintenance value then an evaluation approach based on assessing risks of future decline in these benefits and a policy response to reduce these risks are the appropriate means to measure and manage such "at risk" environmental benefits.

4.   Bibliography

ADGER, W.N. and M.C. Whitby (1993), *Natural resource accounting in the land-use sector:* Theory and practice. European Review of Agricultural Economics, 20. pp. 77-97.

ARROW *et al.* (1993). R*eport of the NOAA Panel on Contingent Valuation,* Federal Register, Vol. 58, No 10, January 15 1993 pp. 4601-4614.

BROMLEY, D.W. (1996), *The Environmental Implications of Agriculture.* OECD Seminar on Environmental Benefits from a Sustainable Agriculture. Helsinki 10-13 September 1996.

CHILTON, S.M. (1996), *Developing Local Community Sustainability Indicators: Lessons from America.* Working Paper, Centre for Rural Studies/Dept. Of Agricultural and Food Economics, The Queen's University of Belfast.

CURTIS, A. and T. De Lacy (1996), *Landcare in Australia: Does it Make a Difference?* Journal of Environmental Management, Vol. 46, No. 2.

European Environmental Agency (1995), *ropes Environment: The Dobris Assessment* Office for Official Publications of the European Communities Luxembourg.

HEIMLICH, R.E. (1995*), Environmental Indicators for US Agriculture* in: (ed.) S.S. Bati.e. "Developing Indicators for Environmental Sustainability": Proceedings of the Resource Policy Consortium Symposium. Washington, D.C.

HODGE, I. (1994), *Rural Amenities: Property rights and Policy Mechanisms,* in OECD (eds). The Contribution of Amenities to Rural Development, Paris, France.

HUTCHINSON, W.G., J. Davis and S.M. Chilton (1995*), Theoretical and Spatial Limits to the Value of Rural Environmental Benefits.* Journal of Rural Studies. Vol. 11 No. 4 pp. 397-404.

JOHANSSON, P.O. (1994), *Characteristics and Valuation of Rural Amenities* in OECD (eds). The Contribution of Amenities to Rural Development, Paris, France.

KING, D.M. and P.R. Crosson (1995), *Indicators of Natural Capital: Lessons from the Upper Mississippi Watershed Accounting Project* in: (ed.) S.S. Batie. P.R. "Developing Indicators for Environmental Sustainability": Proceedings of the Resource Policy Consortium Symposium. Washington, D.C.

KRUTILLA, J.V. (1967), *Conservation Reconsidered,* American Economic Review Vol. 57 pp. 787-796.

MITCHELL, R.C. and R.T. Carson (1984), A Contingent Valuation Estimate of National Freshwater Benefits: Technical Report to the US Environmental Protection Agency (Washington DC, Resources for the Future).

MITCHELL, R.C. and R.T. Carson (1989), *Using Surveys to Value Public Goods: The Contingent valuations Methods.* Resources for the Future, Washington DC.

OECD (1994a), Agricultural Policy Reform: New Approaches, Paris, France.

OECD (1994b), *Environmental Indicators,* Paris, France.

OECD (1996), *Developing OECD Agri-Environmental Indicators* [COM/AGR/CA/ENV/EPOC(96)47].

UK Department of the Environment (1993), *Countryside Survey 1990 Main Report*, London, England.

WEISBROD, B.A. (1964), *Collective Consumption Services of Individual Consumption Goods.* Quarterly Journal of Economics Vol. 78 No. 3 pp. 471-477.

WILLIS, K.G., G.D. Garrod, and C.M. Saunders (1993), *Valuation of The South Downs and Somerset Levels and Moors Environmentally Sensitive Area Landscapes by The General Public.* Research Report to (UK) Ministry of Agriculture, Fisheries and Food (by the Centre for Rural Economy, University of Newcastle Upon Tyne, England).

Annex: Biodiversity and Landscape

Source: *Environmental Indicators – OECD Core Set/Indicators d'environnement – corps centrale de l'OCDE*, Paris 1994.

The protection of biodiversity and landscapes is both a national and an international goal. It involves different levels of protection for different types of land and ecosystems and includes measures to protect areas, ecosystems and species, and to create biosphere reserves representative of different ecosystems. One important measure to protect biodiversity and landscapes is the creation of protected areas.

La protection de la biodiversité et des paysages est un but à la fois national et international. Elle utilise différents niveaux de protection pour différents types de sols et d'écosystèmes et comporte des mesures visant la protection de sites, d'écosystèmes, d'espèces et la création de réserves de la biosphère représentatives des différents écosystèmes. Une mesure de protection importante est la création de zones protégées.

The indicator presented here presents land areas under management categories I to V of the International Union for the Conservation of Nature (IUCN) classification. This classification specifies different levels of restrictions in human activities. Management categories I and II (Scientific Reserves and National Parks) reflect the highest protection level.

L'indicateur proposé ici présente les superficies terrestres appartenant aux catégories de gestion I à V de la classification de l'Union Internationale pour la Conservation de la Nature (UICN). Cette classification précise les différents niveaux de restrictions imposées aux activités humaines. Les catégories de gestion I et II (réserves scientifiques et parcs nationaux) représentent le niveau de protection le plus élevé.

Protected areas change over time: new areas are created, boundaries of existing areas are revised and some sites may be destroyed through industrial development, shifting agriculture or natural disasters. Actual protection levels and related trends are difficult to evaluate as they are not only a matter of the number and area of protected sites but also a question of the effectiveness of management and of the achievement of protection objectives.

Les zones protégées évoluent rapidement : de nouvelles zones sont créées, les limites des zones existantes sont modifiées tandis que d'autres zones sont détruites par le développement industriel, l'agriculture ou les catastrophes naturelles. Il reste difficile d'évaluer les niveaux de protection réels et leur évolution : ce n'est pas simplement le nombre de sites et leur superficie qui compte, mais plutôt l'efficacité de leur gestion et le respect des objectifs fixés.

The graphics on the next pages present trends in protected areas since 1980 as a percentage of total land area, as well as the level of protection by mangement category in 1990. The table presents corresponding data.

Les graphiques ci-après présentent l'évolution des zones protégées depuis 1980 en pourcentage de la superficie totale, ainsi que les niveaux de protection par catégories de gestion pour 1990. Le tableau présente les chiffres correspondants.

When interpreting this information, it should be kept in mind that definitions, although harmonised by the IUCN, still may vary among countries, and that the total area protected does not always reflect the same level of protection.

En interprétant ces informations, il faut tenir compte du fait que les définitions, bien qu'harmonisées par l'UICN, peuvent encore varier d'un pays à l'autre, et que la superficie totale protégée ne reflète pas nécessairement le même niveau de protection.

Major protected areas/principales zones protégées

Scientific reserves and national parks/ réserves scientifiques et parcs nationaux

	(1 000 km²)			% of territory/ % du territoire	(1 000 km²)	
	1980	1985	1990	1990	1990	
Canada	569.6	642.2	701.3	7.0	268.1	Canada
USA	473.9	649.5	983.0	10.5	204.5	Etats-Unis
Japan	21.3	22.0	24.0	6.4	13.1	Japon
Australia	250.7	354.1	456.5	5.9	295.8	Australie
New Zealand	26.2	27.9	28.4	10.6	25.0	N. Zélande
Austria	2.6	3.0	15.9	19.0	-	Autriche
Belgium	-	0.1	0.7	2.4	-	Belgique
Denmark	0.1	1.3	4.2	9.8	-	Danemark
Finland	4.8	8.0	8.1	2.4	5.1	Finlande
France	12.8	16.5	47.8	8.7	2.8	France
Germany	2.9	5.3	29.6 [49.5]	11.9 [13.9]	[0.1]	Allemagne
Greece	0.6	0.6	1.0	0.8	0.5	Grèce
Iceland	7.5	7.9	9.3	9.0	1.8	Islande
Ireland	0.1	0.2	0.3	0.4	0.2	Irlande
Italy	4.1	5.2	13.0	4.3	1.3	Italie
Luxembourg	0.8	1.1	..	..		Luxembourg
Netherlands	1.1	1.6	3.6	9.5	0.1	Pays-Bas
Norway	37.9	47.2	47.6	14.7	45.5	Norvège
Portugal	2.5	3.8	4.5	4.9	1.0	Portugal
Spain	16.8	17.0	35.1	7.0	1.2	Espagne
Sweden	10.6	15.9	17.6	3.9	5.9	Suède
Switzerland	0.2	1.2	1.1	2.7	0.2	Suisse
Turkey	2.3	2.9	2.7	0.3	2.0	Turquie
UK	13.2	15.5	46.4	18.9	-	Royaume-Uni
OECD	1 462.7	1 850.1	2 481.7 [2 501.6]	7.7 [7.8]	[885.4]	OCDE

Source: *Environmental Indicators – OECD Core Set/Indicateurs d'environnement – corps central de l'OCDE*, Paris 1994.

Major protected areas / Principales zones protégées
Other EC countries / autres pays de la CE

% of territory/du territoire

Protection categories (%)/
catégories de protection (%)
1990

% of territory/du territoire

Protection categories (%)/
catégories de protection (%)
1990

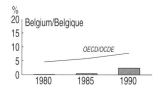

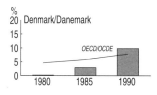

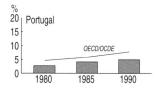

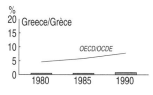

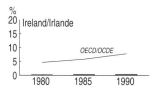

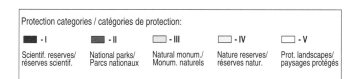

Protection categories / catégories de protection:

■ - I ▨ - II ▧ - III ▢ - IV ▢ - V

Scientif. reserves/
réserves scientif.

National parks/
Parcs nationaux

Natural monum./
Monum. naturels

Nature reserves/
réserves natur.

Prot. landscapes/
paysages protégés

117

Major protected areas / Principales zones protégées
Other Nordic countries / autres pays Nordiques

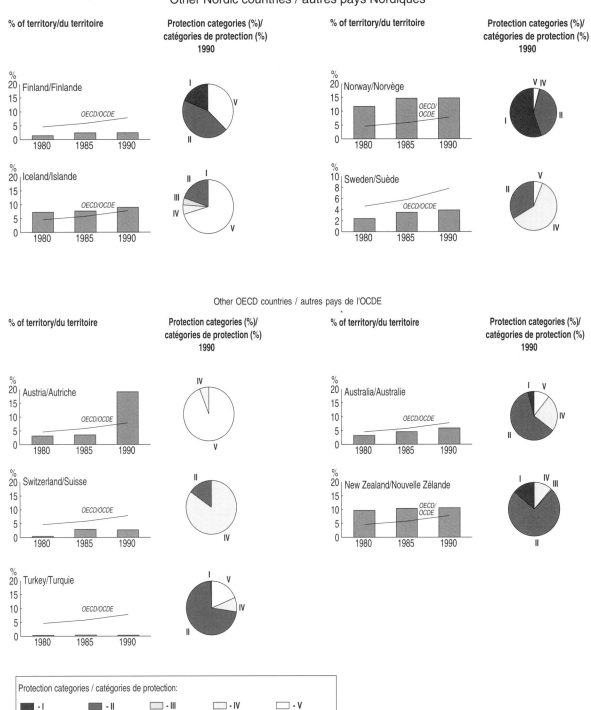

Other OECD countries / autres pays de l'OCDE

Protection categories / catégories de protection:

- I   Scientif. reserves/ réserves scientif.
- II   National parks/ Parcs nationaux
- III   Natural monum./ Monum. naturels
- IV   Nature reserves/ réserves natur.
- V   Prot. landscapes/ paysages protégés

OFFICIAL STATEMENT

by
His Excellency Kalevi Hemila, Finnish Minister of Agriculture

Agriculture affects the environment in many ways. The basic function of agriculture and food production has various kinds of side-effects on waterways, air, soil, landscape and biological systems. Some of these effects are adverse, others inherently carry with them benefits, some of which may not be traded on the market. While the negative effects of agriculture on the ecosystem have received much attention in the past few years, the positive effects of sustainable agriculture have largely remained unobserved. It is not my task to give any more detailed accounts of these benefits at this point; I'll just note that they seem to be connected with the biodiversity in traditional agricultural systems, and the cultural landscape which enhances the aesthetic value of the scenery. It can be asked whether these unpaid services, provided basically by farmers, should be taken into account on the credit side of the record book.

In a country like Finland, of which 70 per cent is covered by forests, 10 per cent by lakes and only 7 per cent by arable land, the open landscape offers a pleasant variety. We usually talk about the benefits of an open landscape in contrast to a pure forest, which in many cases is the alternative to agriculture. An open landscape is characterised by rich flora and fauna that originated in Finland during the warm period following the ice age. Because of the biotopes and niches created by traditional agriculture, many vascular plants were able to survive when the weather turned colder, and spruce trees invaded the country some 1000-2000 years B.C. These plants and biotopes are enjoyed by Finnish citizens who, by the ancient Nordic "every-man's right", can move around freely on privately owned land as well as forests.

These benefits are nothing new. They have always been there but have not received much attention before types of biotopes started disappearing. Now that farmers are facing production restrictions, decreasing producer prices and international competition from other countries we are, however, faced with the possibility that the agricultural landscape will disappear from certain parts of the country. The impoverishment of flora and fauna is to a large extent connected with the disappearance of certain natural agri-ecological systems which are not profitable to maintain. Many farmers seem to be willing to maintain biodiversity, if given the possibility. What we need is some form of agri-environmental support for such ecotypes - a form of support that aims at saving some of the benefits while trying to guarantee that adverse affects are minimised. A precondition for this type of support is that it will not increase agricultural production and will thus not distort agricultural markets, which means that it will not lead to any increases in agricultural surpluses.

I do believe that you have chosen the right country to host the OECD seminar. Finnish nature is pure, we have an extensive agricultural production system and a large comprehensive agri-environmental programme with a participation rate covering almost 90 per cent of agricultural land. In other words, our agriculture is becoming more sustainable.

I warmly welcome you to this seminar and hope that you will have the opportunity to have fruitful discussions on the theme of the seminar: the environmental benefits from agriculture - what they are, how they can be measured, and how future generations can be guaranteed the rights to enjoy these benefits. With these words, I declare the seminar open.

OFFICIAL STATEMENT

by
His Excellency Pekka Haavisto, Finnish Minister of the Environment

There is something of a motivation crisis in Finnish water conservation. In the last couple of decades, communities and industry have substantially reduced their discharges onto the water system, but pollution from agriculture has remained more or less the same. This dispersal (or non-point pollution) is felt and seen mostly in southern Finland. Here the biggest towns filter 80 per cent of the phosphorus and nitrogen out of their sewage and spend about two billion Finnish marks on treatment every year. In addition, agriculture in the south releases about 500 tonnes of nitrogen and 6 000 tonnes of phosphorus into the water in one way or another - without any kind of treatment. If treated this would cost 1.5 billion Finnish marks annually.

In the 1990's, there has been some - although not much - success with reducing discharges of nitrogen and phosphorus from agriculture. Attitudes in farming have been slow to change and many people involved in agriculture do not see the urgency of this matter. One factor here is that Finland does not have any environmental legislation specific to farming, and environmental protection in agriculture has mainly had to rely on counselling and financial incentives.

In autumn 1992, the Ministry of Agriculture and Forestry and the Ministry of the Environment approved a special rural programme designed to reduce harmful environment impacts, and the 1995 budget included environmental subsidies for agriculture for the first time ever. Nearly all Finnish farms - that is, over 80 per cent - receive the basic subsidy and thereby commit themselves to making certain environmental investments. However, this basic subsidy has been available with only a modest input to these investments to these investments from the farmer.

The other type of support is called 'special subsidy', to provide for wide protective strips around fields, develop organic farming methods, take care of heritage landscapes, and encourage the breeding of indigenous animal stock and the intensive use of manure. Very few farms receive this special subsidy. Only some 5 per cent of the country's arable land is covered by it, and in 1996 many farms planning ambitious environmental investments and applying for subsidy have failed to get it. However, the 1997 budget will include larger appropriations for such environmental investments. One of the most popular targets has been organic farming, and about 30 per cent of all special subsidies are used for this purpose.

Organic farming: the farming of the future

Organic farming has grown ten-fold over the last five years. In 1997 it is estimated that some 100 000 hectares of arable land will be farmed organically - that is, 6 per cent of the total. The organic farming organisations aim to extend organic farming to 10 per cent of the total arable area by the year 2002.

The eco-market in farm commodities targeted directly to consumers is still underdeveloped in Finland. According to a Ministry of Trade and Industry committee on eco-exports, Finland's most unexploited potential lies specifically in the food sector - partly because of our unspoiled country-side. Thus we should concentrate on raising quality and value added at the expense of quantity. We are simply unable to produce enough bulk products. On the other hand, our decisions on farming should also take global hunger into account, as Lester R. Brown, the head of the Worldwatch Institute, pointed out when he visited Finland in the spring.

In 1996, eco-exports are expected to account for about one-fifth of our total agricultural exports, which are worth about 130 billion marks. The Ministry of Trade and Industry committee on eco-exports predicts that eco-exports will reach a level equivalent to the present export total within the next fifteen years. In future, agricultural producers who want to compete successfully on the market with the purity of their products are likely to join 'certification chains' which strictly regulate the whole food production process from field to table.

Organic farming is also profitable economically. Denmark has calculated that organically produced foodstuffs yield at least 10 per cent more in net export income than the corresponding products of intensive farming. This is partly because organic farms consume up to 50 per cent less raw materials, externally purchased inputs, and energy.

Organic farming is more than just a cultivation method. It means soft technology, a farm-specific nutrient balance, and considerate treatment for domestic animals - everything that is really traditional and close to the earth, which people would like to see throughout agriculture.

Next year Finland will have a new Nature Conservation Act that will also cover landscape protection. This concept is not just conventional nature conservation. It includes elements created by man. Already, many villages have woken up to the need for landscape management - buildings are being renovated, overgrown woodland is being thinned out, and cattle are grazing in the meadows. Farms play an important role in protecting both landscapes and species. The landscape is less interesting and biodiversity declines when farms take fields out of production and put their machinery in mothballs!

Forests: an essential element in the Finnish countryside

We have a secret weapon in the new Europe, for Finland has a lot of greenery per capita - more forests than in any other European Union country. We should stop complaining about the small size of our farms, because almost all of them also include forests. Including such forest, the average Finnish farm is 50-60 hectares. This figure is exceeded only by Britain with its fields, and Sweden with its fields and forests.

Forests are also the cradle of the Finnish culture and soul. Throughout history, forests have played a significant role in the livelihood and culture of the people. For instance, in Finland (as well as in the other Nordic countries) there is a tradition of public access which enables all citizens to move freely in the forests, pick berries and mushrooms and camp there.

The Finnish forests are home to 700 to 800 hundred endangered species and some centuries-old trees, which we are trying to protect with legislation and by expanding our network of nature reserves. Finland was one of the first countries to draw up a programme for protecting old-growth forests, and this now protects some 300 000 hectares of such woodlands in Finland. About a quarter of these forests are in commercial use.

Small-scale forest ownership has led to various kinds of forest management, and the aims of forest owners managing their own forests are diverse. Merely setting up nature reserves is not enough. We also need a different, softer approach to forestry. Measures such as drainage and forest fertilisation are actually a burden on nature and reduce the forest's biodiversity.

In the future, in some way or another, forest-owners have to give a guarantee that "happy trees" live in their forests. In timber and forest certification, we are lagging behind the other leading countries in this sector. Civic organisations have worked hard to develop the FSC-mark and just two weeks ago a certification committee comprising representatives of ministries, the civil service, as well as NGOs, was set up to investigate opportunities to certify forests.

Today, agriculture in Finland is at a cross-roads. We need a new approach and new, bold ideas to survive and to give way to a new era in agriculture.

OFFICIAL STATEMENT

by
Gérard Viatte, Director for Food, Agriculture and Fisheries, OECD, Paris

I wish first of all to thank our Finnish hosts most warmly for the very active role they have played in preparing, arranging and funding this Seminar. I should also like to thank the eight other financial contributors: Belgium, the European Commission, Germany, Japan, the Netherlands, Norway, Switzerland and the United States. This OECD Seminar on Environmental Benefits from Agriculture in Helsinki could not have been possible without the generosity of these nine donors.

I also wish to thank all the participants on the Steering Group for the Seminar whose support has been most valuable, notably in selecting the main themes for the Seminar. The OECD Secretariat also thanks all countries and consultants who have prepared case studies for this Seminar. Last but not least, I want to express my appreciation to all of you who are here today: your participation augurs well for the coming discussions, and I am sure the Seminar will benefit from your broad range of experience, both national and international. I also make a special mention and welcome to the participation of the environmental NGO's and farmers' professional organisations who will certainly enrich the dialogue on this important area. The considerable interest raised by the theme of the Seminar is a pointer, I hope, to the success of the proceedings on which we are about to embark.

The changing agricultural sector in OECD countries

The need to improve environmental quality in all aspects of our lives has become one of the main policy concerns in all countries. Agriculture is no exception. The impact of agriculture on the environment is a major issue for agricultural policy in OECD countries. Every OECD country recognises the importance of ensuring the sustainable management of natural resources, and many have implemented programmes to that effect. This is part of the wider environmental concerns as expressed at the international level in the June 1992 UNCED Earth Summit at Rio. And the environmental dimension has become a major consideration in designing new agricultural policy measures, in adapting existing measures, and in evaluating agricultural policies.

The agricultural sectors in OECD countries are rapidly changing. Lesser people are employed on fewer farms. There are closer links with the upstream and downstream sectors from agriculture. Advances in technology and the application of science to farming are extremely important. Farmers are increasingly deriving part of their incomes from off farm sources. All of these changes have implications for the use of resources in agriculture, with environmental consequences.

Many farmers and farm business have responded to the increasing demand to adopt innovations in pest, disease and fertiliser management, and to produce "organically" or "environmentally friendly" foods produced on attractive, well managed farms. Farmers are increasingly conscious of adverse publicity by the media and pressure groups of farming practices that are perceived to harm the environment, animal welfare or human health. Moreover, as the country case studies presented in this Seminar show, farmers are also aware of the need to fulfil other functions demanded by the society, including agriculture's contribution to country-side management, the conservation and enhancement of landscape, and opportunities for recreation. Some of these benefits are captured by farmers. Some are not. The Seminar should help to identify the respective roles for markets and policies (and which ones) to address these benefits.

Farmers also have to face environmental regulations and meet standards in many countries, which can have important implications for their costs of production and competitiveness internationally. However, this has to be viewed in the context of high levels of agricultural support, and the lack of enforcement of the Polluter Pays Principle in many countries. Overall, it is therefore important that policy makers take into account both the environmental benefits and environmental harm from agriculture.

The context of the Seminar in overall OECD work

I would now like to make some remarks about the Seminar in the context of OECD work. Although work on agriculture and the environment has been underway in one form or another for some time, the linkages between agricultural activities, policies and the environment were explicitly identified by OECD Agriculture Ministers at their meeting in March 1992 as one of the three key areas for agricultural policy work of the OECD. At the meeting of the OECD Council at Ministerial level in June 1993, Ministers stated their determination to pursue their efforts to promote agricultural reform, within a comprehensive policy framework, which included addressing environmental questions.

I should stress that work on agriculture and the environment in the OECD is a horizontal activity, involving the Agriculture and Environment Directorates, and their respective Committees, and of course the contribution of experts from OECD capitals. I would also note that there are also close ties with the work of the Group on Rural Development, as well as the Trade Directorate. I mention this because one of the strongest assets of work in the OECD is that it can draw on the mutual expertise of a range of relevant disciplines that are crucial for a complete and balanced analysis of policy.

The work in the Agricultural Directorate is centred on the analysis of markets, the monitoring of agricultural policies, and the examination of policy options in the context of agricultural policy reform. OECD Member countries are already engaged in a process of reform of their agricultural policies, which was agreed by OECD Ministers in 1987. However, overall progress has been slow and limited. But the pressures on government budgets, the need to improve economic performance, and improve trade relations have been factors encouraging efforts. While all OECD countries are engaged in reform, the paths and the pace of reform varies. New Zealand stands out as a country which has now experienced a decade of fundamental agricultural policy reform. Reforms in other OECD countries are more recent and less radical. Nonetheless, the successive reforms of the Farm Act since 1985 in the United States and the 1992 CAP reform in the European Union promise to be significant. I should also note that the concern with environmental issues has been an important factor in most countries as part of their reform efforts.

This reform process has been underpinned by the conclusion in December 1993 of the Uruguay Round Agriculture Agreement. This agreement represents a major step forward from a policy point of view. For the first time agriculture is included in an international trade agreement. But it is likely to have only a limited impact on agricultural markets in the near future. The medium-term outlook for agricultural markets will be mainly influenced by other factors, including economic developments in non-OECD countries. However, the agreement is also significant in the present context because environmental programmes are included in the set of so called "green box" measures, which are not subject to support reduction commitments.

Overall, there are three important related results of these agricultural policy reforms: some shift from commodity linked price support to other means of support to farmers; a greater emphasis on payments to farmers to address environmental issues in agriculture; and some reduction in trade distortions. But I repeat again that support still remains high in many countries. All of these factors have consequences for the environment, and are being considered in the OECD work programme on agriculture and the environment.

Turning now to the Environment Directorate, its work seeks to inject an economic dimension into environmental policy and, likewise, an environmental dimension into economic and sectoral policies. As progress in the field of the environment is highly dependent on action taken in other sectors, this work also includes the promotion of policy integration in a range of sectors, including agriculture. The work on the sustainable management of resources, the analysis of economic instruments, the nature of externalities, the links between trade and the environment, and the development of environmental indicators is an important framework for the activities on agriculture and the environment. I would note that the work recently published on "Saving Biological Diversity" and on "Subsidies and Environment" is also relevant.

OECD Environment Ministers met in Paris in 1996. In their Communiqué they referred to agriculture, stressing that "extensive integration of environmental policy with agriculture, forestry and fisheries policies, nationally as well as internationally, remains an important policy objective. Ministers saw scope for significant further progress by adjusting agricultural support policies and encouraging farmers, fishermen and foresters to adopt environmentally sensitive production and harvesting methods. There is also an important role for the wider use of economic instruments, as well as other methods such as education, skill development, and provision of information, in achieving this objective. Ministers were particularly concerned to ensure that there should be a reduction in environmental damage caused by agricultural pesticides and nutrients".

The Seminar and OECD work on agri-environmental issues

The long-term aim of OECD activities in the agri-environment area is to help policy makers to develop approaches and measures, and promote market solutions to best achieve both agricultural and environmental policy objectives in the most economically efficient ways, while recognising the diversity of environmental situations within and between countries.

The need to analyse such policies has been heightened by the process of agricultural policy reform, which has led many OECD governments to explore new means by which income support can be delivered to farmers in the context of structural adjustment, and the growing public concern over the environmental harm caused by some agricultural practices.

The work in agriculture and the environment has progressed on two main fronts, including both qualitative and quantitative aspects. These are: the analysis of agricultural and environmental policies, and policy reform (including the trade aspects), and the development of agri-environmental indicators. These broad areas of work are complementary and intended to:

- develop a better understanding of agri-environmental relationships, especially the links between agricultural policies, farming practices and the environment;

- examine policy options at the national and international levels and explore possible ways for improving agricultural and environmental policy coherence;

- provide OECD countries with an opportunity to share information, data and policy experiences in addressing environmental issues in agriculture.

Some of the results of the work have already been (or are being) published, as for example, on sustainable agriculture, on forestry, agriculture and the environment, on the environmental effects of agricultural land diversion schemes, on co-operative approaches to address environmental issues, and on developing agri-environmental indicators.

I am well aware of the concern in many countries with "food security", and I strongly believe that an efficient market oriented sector is essential to achieve this. Achieving the objectives of increased agricultural output to feed a growing and richer population, while not harming the environment is a challenge. But is possible - if the right signals are given to farmers, and the appropriate balance of policies is implemented. This requires a co-operative and coherent approach to policy.

Another area which is closed linked with agri-environmental issues is rural development. An important aspect of the agricultural policy reform process is to provide farmers and their families with opportunities to find other ways of increasing household incomes, unrelated directly to support for farm production. One way is for farmers to provide rural amenities goods and services. Such amenities are very diverse in nature and are produced by many sectors of the rural economy and not just by farming.

The work of the OECD Rural Development Programme has improved the understanding of the role of amenities and their contribution to the development of rural areas. It is a step forward in the understanding of the nature and valuation of rural non-market goods and services, and appropriate policies, which is also the subject of the present Seminar. The work recently published on "Amenities for Rural Development" is particularly relevant.

There is a further dimension that is important. OECD countries are increasingly inclined to link support to farmers to the environmental goods and services provided by agriculture. Although to date they represent only a minor part of the overall farm support, over the last few years there has been an increase in payments for the provision of "environmental services" in agriculture, for both the physical environment (ecological contributions) and social environment (cultural values). Considerable efforts are being made to devise and introduce agro-environmental measures and programmes, although there are very different approaches (for example, the Landcare programme in Australia, agri-environmental measures in the European Union, the Conservation Reserve Programme in the United States, and the ecological measures as an integral part of agricultural policies in Switzerland). But in many countries these policies are being implemented alongside programmes that give substantial agricultural support, which contributes to generating adverse effects for the environment.

What will the Seminar achieve?

The Seminar is an integral part of the programme of work of the OECD Joint Working Party of the Committee for Agriculture and the Environment Policy Committee (JWP). It offers an opportunity to provide a step forward in the OECD reflection on the provision of environmental benefits from agriculture, as a part of the agricultural reform process, and the better management of natural resources.

The area covered by the Seminar is highly policy relevant. But much of the analysis and understanding of the issues are evolving, and many countries are only at early stages of developing policies and approaches. This is important because for many of the "environmental benefits" from agriculture there may currently be no markets, although some could be marketable. It is hoped that the papers and discussions in the Seminar will advance the understanding in this area. They will provide an important contribution to identifying the key policy messages in the Synthesis Report of the work on the JWP, and to the forthcoming meetings of the Committee for Agriculture and the Environment Policy Committee at Ministerial level in 1997 and 1998.

I return once more to the framework in which the Seminar is taking place. The OECD has agreed to the reform of agricultural policies, and to the sustainable management of natural resources. The results of this Seminar should provide a valuable contribution to the policy debate in the OECD, in a forward-looking perspective. This could be achieved by:

- developing a better understanding of the nature, values and analysis of the environmental benefits from agriculture;

- recognising the need for a balance between environmental benefits and environmental damage, caused by agricultural practices and policies in many countries;

- sharing experience and providing a critical assessment of the policy measures currently used for the enhancement of these benefits through the analysis of country experiences;

- defining the conditions under which environmental benefits could be remunerated through markets drawing on the innovative capacity of farmers;

- helping to identify appropriate, targeted policy measures to complement agricultural policy reform, and the need to take a coherent approach to agricultural and environmental policies;

- outlining the institutional framework, and the role of public and private sectors at all levels - local, national and international - to improve environmental performance.

To conclude, I wish the Seminar all success. I trust that the discussions will be both lively and productive, and that the Seminar will help to provide a number of useful signposts and pointers for the discussion and development of policy relating to environmental benefits from a sustainable agriculture.

SUMMARIES OF THE COUNTRY CASE STUDIES ON POLICY MEASURES AND PRACTICES TO PROVIDE ENVIRONMENTAL BENEFITS OF AGRICULTURE IN OECD COUNTRIES (PREPARED BY THE COUNTRIES THEMSELVES)

1.   European Union: The   European Union's   agri-environmental   Regulation 2078/92 - Framework

While some environmental problems can be addressed through codes of good agricultural practice, backed up where necessary by legal restrictions on some farming techniques, in many cases the enhancement or preservation of the rural environment require a level of commitment well beyond the legal minimum and beyond codes of good agricultural practice. This is particularly the case when the natural biodiversity is dependent on specific types of farming and the associated landscapes which, in Europe, are the product of centuries of agricultural activity. Compared with the modernisation of agricultural techniques and intensification of production which has taken place over the last 50 years, environmentally beneficial measures often entail considerably reduced productivity and costly and time-consuming improvements. For this reason, payments from public funds for farmers who alter their activities substantially to benefit the environment is essential to achieve the environmental goals which society wants.

Recognition of the role of farmers as protectors of the environment and guardians of the countryside is now established policy of the Community. The successful implementation of policies such as the agri-environment programme constitute a substantial part of the European Union's obligations under Agenda 21. Through the agri-environment programmes, farmers agree to undertake efforts to maintain or improve the environment and may be paid *premia* which correspond to the costs incurred and income foregone in adopting the environmentally beneficial techniques. The measures supported fall into four broad categories: low-intensity farming systems; landscape; set-aside and maintenance of abandoned land; and training and demonstration projects. Implementation in the Member States has been very diverse and in some countries take-up has been low, which may indicate that many farmers prefer to stay with intensive systems. Some evaluations have been completed in respect of programmes which had been established before 1992. However, it is too early to make a full assessment of the impact of programmes across the European Union, since the implementation started mainly after 1993. Member States have strategies for monitoring and evaluation designed to measure changes on the environment and reductions in production. A key element in the future success of evaluations will be the continued development of indicators.

2.   European Union: The   European Union's   agri-environmental   Regulation 2078/92 - Examples of environmental benefits

The open, non-forested, component of the European land area, which ecological evidence suggests is more ancient than often realised, together with the long history of farming in Europe, have provided continuity and stability for plants and animals to adapt their often complex lifestyles.

Certain management now maintains spatial and temporal biodiversity, at a variety of scales, on farmland within which a whole range of ecological and behavioural processes are operating. Detailed examples are used in this country case study to describe this in more detail.

Although the ecological conditions on farmland are often complex (involving interactions between management practices, plants, invertebrates, birds and mammals) it is clear that relatively simple adjustments to routine farming operations can provide and maintain farmland high in biodiversity. Policy instruments such a Regulation 2078/92 should, in principle, therefore be capable of achieving real environmental benefits which are both sustainable and cost-effective. Indeed, without an ability to maintain certain styles of farm management across a wide range of farming systems, it will be difficult to conserve many of Europe's most threatened plant and animal communities; or even to meet the European Union's nature conservation objectives within the Wild Birds Directive and the Species and Habitats Directive.

3.   Austria: Organic farming

Organic farming can be defined as an approach to agriculture where the aim is to create integrated, sustainable agricultural production systems, and maximum reliance is placed on self-regulating agro-ecosystems in protection from pests and diseases. It means a management of biological processes and interactions. Organic farmers "feed the soil and not the plant directly". So-called "weed" is seen by organic farmers as "accompanying herbs", tolerated to form a complex unity. Reliance on external inputs is reduced as far as possible. If there is a lack in animal dung as manure, compost-systems are elaborated.

It is a concept of the farm as an organism, in which all the component parts - soil, micro-organisms, insects, plants, animals and humans - interact to create a coherent and stable whole. The major factor which distinguishes organic farming from other approaches to sustainable agriculture, is the existence of strict standards and certification procedures, both legislated and voluntary. So we can define a clear dividing line between organic and other farming systems. As explained in the detailed case study, organic farmers in Austria have followed the EU-regulation 2092/91, since 1994. Besides this regulation, the requirement from Austria's "Codex Alimentarius" still exists, specifically for animal husbandry. This is the concept of whole farm management.

In 1990 the number of organic farms in Austria was only 1 500, but rose constantly to 5 782 holdings in 1992, and around 18 500 holdings in 1995, which is 8 per cent of all farms in Austria. But out of them only 15 844 get support under EU-Regulation 2078/92. This steep increase was possible by improved public support. Many farmers, interested and willing to follow the rigid regulations of organic farming, could not afford to switch earlier to this sustainable form of agriculture. Besides lower crop-yields, the process of transition to organic systems involves the restructuring of the whole farm business and is a complex process, involving a high degree of innovation and learning on the part of the farmer. The public is interested in organic farming because, according to the principles of organic farming, the use of inputs is much less than in conventional farming (except manpower), and less inputs, means less energy-use, and less pollution.

A number of studies show the specific environmental benefits delivered by organic farming. For example, carbon-dioxide-emissions from organic farms are estimated at about 60 per cent less than at conventional farms, due to less energy consumption; groundwater contamination by nitrate-leaching under conventional farms is estimated to be much higher than under organic farms. Concerning biodiversity, if we include micro-organism and the diversified crop rotation, it is clear that the biodiversity is bigger within organic farming systems.

The price for not using conventional inputs is a reduction of crop yields but higher man-power involvement. Concerning yields, a recent comparison in Austria (1995) shows that the difference is quite significant for cereals, and potatoes, but less marked for peas and sunflower. The lower quantity of outputs is not always compensated by lower input costs (manpower increase), and there is no guarantee that "organic products" fetch automatically better consumer prices. Unfortunately, consumers do not distinguish between real nourishment and simple provision. Although the portion of an average income spent for food, dropped to 17.2 per cent in Austria, (only 20 per cent of this portion comes to the farmer), consumers are always attracted by cheap foodstuff. It is clear that for the survival of any wanted type of agriculture, the public/consumer has to support it. That would be the price for the rendered service.

4.   Finland: Environmental benefits from agriculture

The role of agriculture in the Finnish society is changing. Agriculture has gained more relevance as a maintainer of the landscape and preserver of biodiversity than as a producer of milk, meat and cereals. Agriculture has both positive and negative impacts on the environment. Agricultural landscape for all to see and enjoy can be seen as a positive externality of agricultural production, and the nutrient load on waters can be regarded as a negative one. The greatest threat to rural landscapes are caused by discontinuing cultivation, depopulation of rural areas and the closing of the open cultivated landscape. Open landscape is the main characteristic of rural countryside in Finland.

Membership in the European Union brought about the task of adapting the European Union's environmental support system to Finland. The Finnish Agri-Environmental Program (FAEP) was formulated in this process to fulfil the environmental guidelines and recommendations set for agriculture during the past years. It is also seen as a means to promote the role of agriculture as a producer of such public goods as landscape and biodiversity. The FAEP includes the General Agricultural Environmental Protection Scheme (GAEPS) and the Supplementary Protection Scheme (SPS), as well as schemes for advisory services and training, and for demonstration projects. The GAEPS is targeted to all farmers, but the SPS is targeted to environmentally more narrow or restricted local or regional measures. The GAEPS can be seen as a means to decrease environmental degradation, and the SPS as a way to further enhance the quality of the environment and to produce positive externalities.

The first year of implementation shows that the number of farmers participating in the programme is higher than was anticipated, and positive environmental impacts can be expected in the near future. Variation in participation between the support areas and production types was quite low. However, there exists a decreasing trend in participation from the north, especially in the case of pig and poultry farms' participation in the GAEPS which was lower in 1996. Participation of farms with grazing cattle was high, and the most important reason for their non-participation was the investment need for manure facilities. The high rate of participation of animal husbandry farms is expected to result in reduced nutrient leaching from manure spreading and manure storing facilities. However, some animal husbandry farms might leave the GAEPS or stop the production if the investment need for manure facilities is too high and no investment aids are available. Farmers' willingness to join the

GAEPS, even when it might cause them additional costs, also signals their willingness to contribute to improving the rural environment.

Crop farms participated in the GAEPS on a very large scale and in one of the areas of the programme, the GAEPS covers the cultivated area almost totally. Areas of intensive crop production are covered by the GAEPS in a very large scale, which can be expected to result in a positive development in solving the agri- environmental problems of those areas. Perhaps the most important single change in farming practices due to GAEPS criteria was setting up some 28 000 km of buffer zones by water routes and main ditches, which will have a positive contribution in decreasing erosion and nutrient leaching. It must be noted that 1995 was the first year of the GAEPS and the criteria will not have a full impact until 1996 and 1997, when all the criteria must be fulfilled. In 1995, approximately 82 per cent of the farmers applied for GAEPS premiums. This means that 90 per cent of the arable land is cultivated according to the criteria set in the programme, and 96 per cent of arable land in one of the areas of the programme. The GAEPS will have a positive impact on the environment as fertilizing levels decrease and nutrient runoffs will be reduced. However, these environmental impacts can only been seen over a period of time.

SPS measures aim at producing further positive environmental impacts, such as increasing biodiversity and maintaining the landscape. With these measures, organic farming became a very attractive option for farmers, and is now an economically competitive alternative to conventional farming, especially in dairy production. Economic impacts of the FAEP are still unclear, but Finnish agriculture will move towards more extensive production as income from the sales of the products will form a smaller share of the total farm income. Less intensive production, buffer strips, better nutrient and pesticide management, organic farming; all these will contribute to a more diversified countryside and rural landscape. However, it is too early to say how positive the impact of the FAEP on the environment will be in the long-run. For instance, verifying the decreased nutrient leaching and its positive impacts will take several years, but positive environmental impacts can be expected in the near future.

5. France: The value of aid to mountain and hill farming from the standpoint of balanced land-use planning and development

Agriculture's environmental benefits are not confined to the conservation of endangered natural resources, the maintenance of ecosystems of scientific interest and the preservation of landscape. In a broader sense, they can also encompass a number of benefits to society that stem from the fact that farming activities are kept alive in areas where physical conditions are highly adverse. Maintaining agriculture's contribution to balanced land-use planning is therefore an important agricultural policy goal in many OECD countries and the aid to mountain and hill farming makes an interesting case study. In order to cope with denser population in very confined urban areas, and the attendant social and environmental problems, policies of aid to mountain and hill farming in France are aimed at: developing areas of low population density in parallel with the development of heavily populated areas (large metropolitan areas); and preserving some social diversity and, in particular, individually run family farms.

These aid policies do more than just limit gaps in income and living conditions between town and country (or mountain) areas. Country and mountain areas play a vital role in fostering harmonious development in France; rural environments offer society opportunities for integration, social innovation and another way of life which satisfy important needs. The interactions among agriculture, the natural environment and the needs of urban populations are not always perceived consciously, but they clearly exist. While aid to mountain farming has not fully prevented the drift from the land, it has nevertheless slowed it down and restored a degree of confidence to farmers in those areas. One frequently quoted example in France is the Beaufort region in the Alps; farmers there have done much to develop the region by promoting their cheese and contributing to the expansion of tourism by enhancing the landscape and local amenities. Today, however, this kind of success is only possible with very substantial aid from the community.

6.   France: Sustainable development plans

An experiment was conducted between 1993 and 1995 involving 1 200 farmers in 59 small regions throughout French soils. This was an innovative experimental project, where the objective was to set up a system of economically stable, sustainable production systems, to ensure a quality of the environment and contribute to local development. The project was supported by the European Union, French authorities, professional agricultural authorities, the National Association for agricultural development, and various other associations. The main criteria of the project were: (i) to form groups of voluntary farmers who will commit themselves to the proposed approaches; (ii) to associate environmental and rural development partners in the implementation of the project, setting up technical committees and steering group committees, on a local as well as a national level; (iii) to ensure national coherence by establishing guidelines, common plans of intervention within a national effort.

The ways in which the project was implemented at the sub-regional and farmland levels are steps towards a global approach, which takes into account at the same time technical-economical, environmental and social aspects. The project comprised various parts: (i) study of territory/landscape, to define the advantages and disadvantages of each site; (ii) agri-environmental study of each farm; (iii) development of scenarios for the evolution of agricultural production systems; (iv) contracting of projects between the farmer, the government and the different partners.

The outcome of contracting these projects will be available in 1997, and the results will serve as a basis for setting up new sustainable development projects in 1998, according to the interest expressed since 1996 by the Minister of Agriculture. At the same time, an experimental programme is in progress to integrate sustainable agriculture into training programmes in agricultural teaching.

7.   Germany: Promotion of positive environmental impacts of agriculture

Some programmes are being offered to everyone, as of the beginning of 1995, to promote farming methods in accordance with environmental protection requirements under Regulation (EEC) 2078/92 of the Länder in the Federal Republic of Germany. In setting up programmes according to article 3, Länder governments more or less referred to the specific shape of the "old" EC-extensification programme under their responsibility and to their own particular programmes for nature preservation. During the first five-years' period 1993-97, there will be a total financial volume (EU; federal- and Länder shares) of up to DM 3.5 billion, averaging up to DM 700 million per year. According to available information, in 1995 in the Federal Republic of Germany, applications for promotion according to Regulation (EEC) 2078/92 have been made for a total area of about 5 million hectares, about one-third of agriculturally used land.

Programmes put forward by the Länder in implementing Regulation (EEC) 2078/92 show considerable differences with respect to the number and types of measures included as well as to promotion conditions, particularly regarding the size of the premiums offered. The example of promotion of organic farming as a horizontally offered programme had shown differences between Länder which cannot totally be ascribed to interregional differences in opportunity costs due to locational and structural factors. These differences must rather be attributed to other additional determinants, among which differences in estimating impacts of programme participation on farmers' incomes and also differing priorities for the use of scarce public funds may be of some significance.

From an economic as well as from an ecological point of view the "action-" or "input-oriented" way of promotion should be compared with the "result-" or "output-oriented" promotion measures. This aspect of the application of Regulation (EEC) 2078/92 is which mentioned in this country study. The arguments presented above may have made it quite clear that a complete substitution of the presently applied "action-oriented" type of promotion would not be practicable and therefore not be advisable. But it seems possible to supplement the promotion systems presently in practice by certain elements of a "result-oriented" promotion in order to improve their efficiency.

8.   Germany: Contributions of agriculture to the conservation of biodiversity in cultivated landscapes - possibilities - limits - prospects

Since the Conference of Rio, the conservation and promotion of species diversity, the so-called biodiversity - have been recognised as a global ecological challenge. Comprehensive strategies for nature conservation should be aimed at the preservation and promotion of vegetation types and species typical of a region - not only rare and endangered species - as a genetic resource and as the basis for the conservation of intact ecosystems (protection of biotic resources). Strategies also should be aimed at the protection of soils, water and air (protection of abiotic) to guarantee the conservation of the habitats of all species of a natural area. Agriculture and forestry play a decisive role in the realisation of these aims since they occupy between 60 and 90 per cent of the surface area in the cultivated landscapes of Central Europe.

The development from a natural landscape predominately covered with forest to a more or less open cultivated landscape over the past 6 000 years until approximately 1800, has led to an increase in the number of species of plants and animals. This increase resulted from various, mainly extensive land uses which created completely new types of biotopes, namely meadows, pastures, heaths and fields (so called open-biotopes); plants and animals from neighbouring regions were able to migrate to these anthropogenic biotopes. Despite the increase in the number of species, there also appeared a number of environmental problems, like over-utilisation, strong decline in forests, erosion etc. over the past centuries. With the development and use of open biotopes, evolutionary biological processes had also been triggered off simultaneously, which led to the emergence of new species. Thus, the predominant number of species which newly emerged over the past 6 000 years in the Central European area are tied to open biotopes. Obviously, habitat factors have a stronger effect on plants and animals in open, unshaded sites than on forested ones.

Over the past fifty years, the system-inherent contribution of historic agriculture to the conservation of species and biotopes had inevitably declined to the same degree as the intensification of agriculture has increased. Since the 1970's, the adverse effects of the intensification of agriculture, e.g. nitrate and biocides, problems in soils, waters and ground water or eutrophication of the landscape, have become increasingly visible. A reduction of species diversity occurred, manifesting itself visibly in the "Red Lists of endangered plant and animal species". In this context it has to be stated that nitrogen can change natural as well as anthropogenic ecosystems in a more far-reaching manner than all other environmental factors. Many plant and animal species of open biotopes which have become rare or endangered have small chances of survival under the influence of high nitrogen levels.

One of the most urgent tasks of nature conservation concerning the few natural and semi-natural ecosystems is to secure their continued existence, increase their share and keep them entirely or partly, free from use. In the category of extensively used ecosystems of the historical cultivated landscape the highest efficiency with regard to the conservation of species and biotopes will be ensured, if the existing extensive uses are safeguarded in good time and on a long-term basis (by adequate remuneration as an ecological service, i.e. programmes of nature conservation and by closely tying extensive use to local farms).

In the category of current agroecological systems, experience with the field margin programme has shown that a high efficiency on herbicide-free margins can always be noted when a high natural seed supply still exists in the soil and less especially when productive fields are concerned. Another example of current agroecological systems are extensification schemes on grassland which are often viewed favourably with regard to their importance for species and biotope conservation. Here it has to be taken into account that the initial trophic level and the respective soil type are of decisive importance. For example, if soils have a high nitrogen level, a more or less phase of reduction of nitrogen will be required, before ecosystems can regain suitable living conditions.

If agriculture in Germany - as well as in other countries of the European Union - were to play not merely a marginal role as at present, but an essential role in the conservation of biodiversity, the corresponding framework conditions would have to be created. This calls above all for the acceptance of the conservation and the promotion of the still existing species diversity as an ecological service and its adequate rewarding. The programmes for nature conservation and cultivated landscapes of the EU-states are a first step, but they frequently require further development and financial support.

9.   Greece: Policy measures and practices and environmental benefits from agriculture

Greece is no doubt, the oldest evidence of the development of the European agrarian systems. The model of economic growth that has influenced the present day agricultural commodity market, has led to the creation of only certain "pockets" of intense farming, mainly following the most easily accessible flat and coastal routes of the continental part of the country. Thus the inaccessible mountainous and insular regions were left in the shadow of this development process. It is for this reason that there still exists today, two distinct models of production, namely the intensive farming system and the traditional extensive one.

The degree of intensification in Greece is quite limited. This is mainly due to the geomorphological structure as well as to the small scale farming that prevails in the country. Practically, all plantations are interrupted and surrounded by vast quasi natural areas (hills, ravines, mountains, lakes, gulfs etc.) that are beneficial to preserving biodiversity and to forming interesting and important cultural landscapes. Even the largest plains are extremely small compared with northern European plains. Due to these facts modernisation and intensification of agricultural practices have not yet resulted into severe negative implications and degradation of the traditional local ecosystems.

On the other hand, the traditional extensive farming systems are spread all over the mountainous and sloping insular areas of the country and they are mainly due to a backwardness in the development process. The main environmental benefit in Greece arises from the very existence of those extensive farming systems. However, due to their low competitiveness, such high natural value farming systems are under pressure, either to get totally abandoned or to change into more intensive and hence less sustainable ones, or even to change into non-agricultural land uses.

Retaining most of the highly productive land to agricultural purposes may result in a relatively higher environmental benefit than would have been the case if this land was used for other non-agricultural purposes. On the contrary, abandonment of agricultural activity leads to a gradual environmental deterioration (destruction of land terraces, to water run-offs and soil erosion - almost 30 per cent of the land is under erosion - loss of insufficient water resources, loss of landscape amenities, loss of biodiversity) and to a spiral that results in further depopulation of rural areas.

Greek experience has shown that, market alone is not in a position to provide enough incentives to farmers to continue their activity under adverse economic conditions that extensive agriculture may dictate. Therefore, agricultural policies that offer incentives for the continuation of agricultural activity and prevents abandonment constitute a critical issue that equally influences all environmental and economic parameters in the area. Centuries-old farming systems may be abandoned or changed "overnight" in response to internal or external forces, including market developments and farm policies. Environmental benefits linked to these extensive farming systems may become irreversibly damaged in this process.

This country study presents four concrete examples, showing how market forces and supporting measures designed at national or even regional level may lead to contradictory results. The first example refers to a fertile area of Greece, the Argolid valley, where farm policies have encouraged agricultural activities with negative environmental effects. However, there is no expectation that the situation could change under market conditions alone. The second example refers to a remote Aegean island, Lemnos, where viniculture consists a major human activity, which has been developed along its long history. The wine produced there is of high quality and its method of cultivation greatly protects the environment. However, farm policy did not prove able to give enough incentives to the farmer to continue working on this land, but market conditions alone could not ensure that vinicultural activity may continue effectively.

The third example refers to how olive groves, cultivated on sloping marginal land, and using highly extensive farming practices, offers benefit to the environment. Olive groves constitute an "artificial" ecosystem which is very close to natural ecosystems, even when olive trees are productive. Especially on remote islands of the Aegean Sea, retaining olive plantations is an imperative both for economic and social reasons, but also for environmental reasons. In fact, it would be a great loss for the environment if an olive grove is uprooted due to its low economic effectiveness. In most cases, it would be better to leave an olive grove to become a "natural" ecosystem instead of uprooting it for the sake of a reduction in the support to agriculture.

The case of another Aegean island, Naxos, is the fourth example used to show how environmental benefits from agriculture can be provided when integrated policies, measures and practices are used to address problems of specific areas. By constructing over 140 small dams in a 3-year programme, the main target was to achieve water management and, through it, to face several agroenvironmental problems of soil erosion, water supply and irrigation. The integrated character of the programme managed to further restore the old agricultural terraces, hence traditional crops have again started to be produced for local consumption. It additionally achieved a certain degree of natural afforestation, without spending water resources.

Finally, it remains a duty for the agricultural policies to find an efficient way to reconcile market reality with environmental benefits from agriculture. In this context, three points should be stressed: (i) environmental benefits from Greek Agriculture are still present, but they will not survive for long unless they are directly related to integrated policy measures that are tailored for specific areas, population groups and activities; (ii) if traditional, extensive farming practices, as well as traditional activities that provide environmental benefits are to be maintained, it is important to recognise that some remuneration to farmers is needed which will not be in contradiction to the general agricultural policy issues; (iii) specifically targeted policies and measures should address the need to maintain the very specific land management techniques that have persisted along the years.

10.   Netherlands: The experience with nature policies regarding farmland

In 1975 the Dutch government published a policy paper on the relation between agriculture, nature and landscape conservation, the so-called Policy Document on Agriculture and Nature Conservation. This Policy Document describes the tensions between the interests of agriculture on the one hand, and nature and landscape conservation on the other, and defines national policies to eliminate these tensions by means of management agreements and land acquisition. The policy is to be carried out on 200 000 hectares of agricultural land with value for nature conservation. This is about 10 per cent of all cultivated land.

In the Policy Document two kinds of areas are distinguished: the 'management areas' and the 'reserve areas'. The policy for the management areas (100 000 hectares) is to reconcile the requirements of commercial farming with the objectives of nature and landscape conservation by means of management agreements. The management that is desirable in reserve areas (100 000 hectares) is incompatible with profitable farming in the Dutch situation. Therefore these areas will be withdrawn from agriculture and their management will be transferred to a nature conservation body. Agricultural land is designated as a management area or a reserve area if it has (potential) value for nature and landscape conservation. The four main categories of valuable semi-natural landscapes are semi-natural grasslands, breeding areas for meadow birds, wintering areas for water birds, and landscapes rich in natural features as hedges, plots of woodland etc.

Participation in the programme of the Policy Document is voluntary, and until 1995, about 125 000 hectares of agricultural land have been designated as management or reserve areas. In these designated areas about 6 000 farmers have management agreements covering about 40 000 hectares. About 32 000 hectares reserve area has been acquired, of which about 10 000 hectares is agricultural land. The Policy Document has been worked out in the Regulation on Management Agreements and Nature Development, which is the national implementation of Regulation (EEC) 2328/91, regarding the Less Favoured Areas (LFA) and Agri-environmental Regulation (EEC) 2078/92. In 1995 the Dutch government paid Dfl 9.0 million in compensatory payments in LFA's of which Dfl 1.9 million was financed by the European Union and Dfl 16.2 million for management agreements of which Dfl 5.4 million was financed by the EU.

To establish the effects on nature, a monitoring programme has been carried out since 1986. Several representative plots all over the country are monitored every three years for meadow birds, and every six years for vegetation. It is too early to present the definite conclusions, but a tendency can be shown. Regarding the vegetation, a management agreement (especially with far-reaching measures) can stop the decline of plant species diversity of grasslands. The remaining valuable botanical grasslands can thus be conserved. However, most of the designated grasslands are now species-poor. In these situations a management agreement with far-reaching measures will give a certain increase in species. The results of the management will be more successful under promising abiotic conditions (e.g. grasslands on chalk or grasslands on seepage-soils). If the abiotic conditions are less promising (e.g. dry sand and clay soils), management on the basis of a management agreement will lead to lightly fertilized grassland types.

Regarding meadow birds, it can be noted that in general most species can be protected with management agreements. The Black tailed-godwit (*Limosa limosa*), Northern Shoveler (*Anas clypeata*), Garganey (*Anas querquedula*) and Redshank (*Tringa totanus*) are doing well in areas where management agreements have been made. More vulnerable species, like Common Snipe (*Gallinago gallinago*), Ruff (*Philomachus pugnax*), Corncrake (*Crex crex*) and Eurasian Skylark (*Alauda arvensis*), are following the national trend, thus also declining in areas where management agreements have been made. It is important to note that this is a tendency and not a definite conclusion. The period of monitoring was too short in most areas to draw definite conclusions.

The increase in the area under management agreements is satisfying. On the other hand the programme is as yet to come into effect on 50 per cent of the designated areas. Therefore it is important to improve the programme in order to get more farmers interested in nature-oriented agricultural management. Improvement of the programme implies that the quality of the management (in terms of nature results) at least remains the same and that participation of farmers grows. In this context, four new initiatives have been discussed: nature result payment, margin management, cloak areas and nature conservation by private landowners.

11. Sweden: Incentives for environmental management of farmland with high biological values

Since the middle of the 1980's, Sweden has had a programme to reduce nutrient leakage and another programme to reduce the use of pesticides and the risks connected to them. The programme includes regulations on the use of pesticides, taxes on fertilisers and pesticides, research and development, information and monitoring. Moreover, certain biotopes are protected by national legislation, which also states that agriculture must take the environment and cultural heritage into consideration.

Different cultivation methods combined with various natural conditions have created a varied landscape with high biological diversity. Structural transformation of agriculture and abandonment of land have led to a decreased variation in the landscape. In plain lands the landscape has become more uniform. About 20 per cent of the species in the Swedish agricultural landscape are considered to be more or less acutely threatened by extinction. In the last decades, overgrowing and/or afforestation has been the most serious threat to biodiversity. Since last year, Sweden has a management agreement programme to conserve biodiversity in the agricultural landscape.

The framework for the programme is Council Regulation (EEC) 2078/92 on agricultural production methods compatible with the requirements of the protection of the environment and the maintenance of the countryside. The overall objectives for the Swedish programme are to maintain a naturally and culturally valuable and varied agricultural landscape, to conserve the biological diversity and to minimise the negative effects on the environment caused by nutrient leakage and the use of pesticides. The programme consists of three parts: the conservation of biodiversity and cultural heritage values (120 MECU/year); environmentally sensitive areas (10 MECU/year); and the promotion of organic production (30 MECU/year).

Farmers have to send in a special application form to receive the compensation. The conditions they have to fulfil are quite extensive and detailed, and normally concern the biological quality and the management of the land. Farms with biologically and culturally valuable landscape elements can get a compensation for the management of these elements. The programme is supplemented by an information campaign, which includes courses, demonstration farms and individual management plans for the natural and cultural values on the farms. The terms for the management agreements are quite clear and detailed as are the conditions for joining the programme. The participation in the programme has so far been very good, but the programme should be evaluated in terms of its cost-effectiveness.

12.   United Kingdom: English environmentally sensitive areas

United Kingdom policy for agriculture and the environment seeks to balance the demand for efficiently produced food with the demand for the countryside to be protected and cared for. Farmers and other land managers are seen to have the greatest role in caring for the countryside, and government policy assists them to reconcile agricultural and environmental needs through a combination of guidance, protection measures and incentives.

Environmentally Sensitive Areas (ESAs) are a policy operated to maintain, protect and enhance the wildlife, landscape and historic environmental value of designated areas, through the encouragement of appropriate agricultural practices. The environmental interest of ESAs has largely been created through traditional agricultural practices which have led to co-evolution of wildlife communities and traditional agriculture, and also to the creation of distinctive local landscapes.

The ESA scheme was introduced in England in 1987; ESAs currently cover 10 per cent of England, and 15 per cent of the United Kingdom as a whole. ESAs were also introduced in Scotland, Wales and Northern Ireland, and the Agriculture Departments to these countries have the responsibility for designating and operating their ESAs. The general principle of the scheme is the same in these countries, although the mechanism has been adapted to deal with local conditions and environmental priorities.

The main distinguishing characteristics of the scheme are that it is based on 5 or 10 year contracts with farmers to carry out a defined set of beneficial agricultural practices in return for fixed annual area-based payments, and is voluntary and open to all managing suitable land within targeted areas. Land managers will not normally be funded to carry out activities which would be expected of them in any case, such a following Good Agricultural Practice. Regulation is used to ensure compliance with such expectations where necessary. Incentives such as those offered under the ESA scheme are used where the environmental goals sought, go beyond those that can reasonably be expected of farmers to carry out of their own accord.

Examples of the results of recent environmental and economic evaluations of the impact of the policy in the first five ESAs introduced in 1987 are described, which illustrate how the scheme is targeted, and specific objectives set and evaluated. This review of the ESA scheme was informed by an extensive environmental monitoring programme. This work enables the success of the scheme to be evaluated against the detailed environmental objectives which have been published for each ESA.

Other economic research into the ESA policy has been carried out. For example, a contingent valuation cost-benefit analysis conducted in 1993 showed that the South Downs and Somerset Levels and Moors ESAs represented concluded that the two ESAs were good value for money: the benefits of the South Downs ESA were shown to be at least five times the cost of payments to farmers in the scheme; in the Somerset Levels and Moors the benefits were at least twice the cost.

The results of environmental and economic evaluations were published during the review. The evaluation confirmed that the ESA scheme was delivering positive environmental benefits which would not have been achieved in its absence. In particular, the economic evaluation demonstrated that farmers were incurring significant opportunity costs through their participation in the ESA. It also showed that whilst most farmers were committed to this scheme in the long term and saw the ESA as a valuable means of environmental protection, many of the practices encouraged by the ESA scheme would not be continued in its absence.

13.   Norway: Environmental benefits by agriculture landscape

The geographical location, the harsh climate and the topography of Norway, limits the possibilities for agricultural production. The farmland is divided into scattered and relatively small plots by lakes, forests, and mountains, but farm structure is also a result of strong historical and cultural traditions.   While sparse population makes many areas vulnerable for depopulation, settlement in all regions is important for the utilisation of natural resources, regional policy concerns, and defence reasons. In order to achieve a sustainable utilisation of land area and natural resources it is a national aim to maintain the overall settlement pattern, and to maintain and develop social structures and services for the rural population. It is recognised that the agricultural sector will be a main contributor in this context. As a consequence, it has been important to secure existing farm-land and maintain an active farming throughout the country.   A comprehensive support system for agriculture has therefore been established.

Farming is important for the preservation and development of the cultural landscape. The agricultural landscape is an object of a perpetual evolution due to changes in factors and processes of production, and environmental state. In a country dominated by forests and mountainous areas, farm land is demanded because of its contribution to variation in the scenery and biodiversity. Since the single elements that contribute to the landscape tend to be viewed as elements of the society's total capital stock, the preservation and conservation of these are highlighted. However, the overall landscape is just as important as the elements. In addition the agricultural/cultural landscape provides the basis for recreational activities.

The agricultural landscape which appears today, is partly a result of the agricultural policy adopted by the Parliament in 1975. An important element in that policy was the "escalation plan" which aimed at an optimal utilisation of the national resources: cereal production was promoted in the south-east areas of Norway, while more labour-intensive animal production was promoted in the remote and less favoured areas, where grass based livestock production is to a large extent the only alternative for agricultural production in these areas. In 1993 the Parliament set new guidelines for the agricultural policy, with the overall objective of creating a more market oriented agriculture. It gave more emphasis to the development of a sustainable production and consumption, high priority to the regional policy aspect of agriculture and granted funds to develop new farm-related economic activities. The focus on sustainable production and consumption made it necessary to shift from reacting on environmental damage to prevent it.

The need for increased attention on the cultural landscape can be explained from environmental and production challenges and international agreements. It can also be found in the increased demand for recreation and leisure activities and new industries in rural areas. The cultural landscape can be considered as a public good like many other environmental benefits. Because of market imperfections its value cannot be entirely determined by ordinary market prices. Different measures have therefore been established to promote a more environmentally friendly development.

The Norwegian policy is founded on a broad range of legislative, economic, and administrative measures, based on voluntariness and commitments from farmers more than on compulsory measures. The measures are targeted towards specific purposes to promote cost-effective actions. The most important general programme is the Acreage and Cultural Landscape Scheme introduced in the late 80's, which provides a direct payment to farmers under certain obligations. In addition, specific measures are used to extend support to landscape maintenance and development; and offer grants for restoration of listed buildings; for amended soil management and for improvement of environmental-technical facilities. The implementation of this set of measures necessitates an extensive administration and control systems to ensure that the policy is carried out efficiently. The public agricultural extension service is responsible for implementing agricultural policy at the local level.

As agriculture plays an important role in preservation and development of the cultural landscape, and contributes to employment and income in rural areas, a local understanding and knowledge is demanded. Interest, commitment, and leadership coming up from the local level by voluntariness and "self-government" are important ingredients in the policy implementation, although it is also necessary to have some national priorities. National and regional plans are needed to give strategies, priorities, and guidelines. It is recognised that the cultural landscape is shaped by many different elements, processes, and activities. In policy making it has therefore been necessary to deal with the whole spectre of problems, causes, and effects by using a mixture of different measures to find comprehensive solutions.

14.   Switzerland: Government policy to stimulate environmental benefits

In June 1996, the Swiss people ratified a new article in their Constitution setting forth the tasks of agriculture: (i) supplying the population with guaranteed food supplies; (ii) preserving the natural bases of existence and maintaining the countryside; (iii) keeping the population geographically decentralised. The new article also requires production to be both sustainable and market-oriented, acknowledges that agriculture is multifunctional, and provides for compensation in the form of payment for services rendered, on condition that it complies with environmental requirements. Therefore, the new farm policy promotes agriculture that is sustainable from an economic, social and environmental standpoint. With regard to sustainability, its objectives are to limit environmental costs and optimise environmental benefits, including: (i) safeguarding the countryside, villages, buildings and pathways; (ii) preserving the natural bases of existence, i.e. water, air and land; (iii) preserving the biological diversity of rural areas; (iv) recycling waste.

Switzerland's agri-environmental policy uses three tools, in descending order of priority. First, research, training and extension are to heighten farmers' awareness of environmental problems and propose solutions thereto. Second, there are financial incentives to safeguard the environmental benefits associated with agriculture. Third, requirements and compulsory charges may be imposed. The main incentive for minimising adverse effects and encouraging beneficial effects on the environment is a voluntary scheme to promote production systems that are particularly environment-friendly (integrated production, organic farming) by giving grants to farmers who comply with the requirements set for the farm as a whole. By giving national recognition to specially designated forms of production, the Confederation can promote sales of such produce.

To keep the Swiss population geographically decentralised and conserve the traditional rural landscape, agricultural policy strives to maintain an agricultural system based on small, individual farms. The purpose of modern land improvements is to create not only optimal structures that will meet the objectives of sustainable development, but also a context in which market-oriented agriculture, environmental protection and land use can flourish. Concerned about the abandonment of certain marginal areas, particularly in the mountains, and the repercussions on their landscape value and the protection they afford against natural disasters, the Swiss government keeps extremely close track of farming conditions and the hardships that face people living in such areas. Stimulating the economy by subsidising new infrastructure and granting aid to mountain farmers are two ways of slowing down rural depopulation.

Agriculture plays a crucial role in preserving biological diversity in the countryside. By encouraging the creation and upkeep of semi-natural environments such as hedgerows, copses, traditional standard-tree orchards, extensive pasture and litter meadows, agricultural policy helps to preserve biological diversity. The Swiss are sensitive to the quality of their landscape which is part of their cultural heritage. For its 700th anniversary, Switzerland has set up a fund for the conservation and management of traditional rural landscapes. This is available for repairs to irrigation channels, shingled roofs, dry stone walls, etc. In addition, as part of its waste policy, the Swiss government has developed schemes for waste collection at source and quality guarantees for the sewage sludge and composts that enable waste to be recycled in agriculture as either fertilizer or fodder. All these economic incentives have well defined objectives. To measure their effectiveness, the government has developed control and evaluation programmes. Once available, the findings should enable Switzerland to justify its agricultural policy at home and abroad and, if need be, promote further adjustment to the incentives to allow for advances in research and for future policy requirements.

15.   United States: The role of agriculture in protecting biological diversity

The country study was prepared by representatives of The Nature Conservancy (TNC), an international, non-profit, nongovernmental, conservation organisation based in the United States. TNC has agencies in all fifty states and works in partnership with other nongovernmental organisations throughout Latin America and the Pacific Basin. The mission of TNC is to preserve plants, animals and natural communities that represent the diversity of life on Earth by protecting the land and water that is needed for species survival. TNC's efforts are largely dedicated to the protection of intact, natural environments, and it could be assumed that agriculture in the United States is in direct conflict with the goals of the TNC. This is definitely not the case where the objective is to protect aquatic systems - where agricultural land use is preferable to residential development. In vast areas of the American West, TNC is working to promote sustainable grazing practices that can restore and protect biological diversity on rangeland.

The country study presents four case studies where TNC is working in partnership with ranching and farming to restore and protect biological diversity. In all cases, the public sector also plays an important role. TNC is protecting the North Landing River around the City of Virginia Beach, which is a rapidly expanding community where development threatens to consume prime agricultural land. If the agricultural area is converted to residences, management of the river's riparian area will be extremely difficult, in addition, housing development poses a serious threat to the water quality of the river. In collaboration with other non-profit organisations and the City of Virginia Beach, TNC has helped create a "Purchase of Development" rights programme for the agricultural land that is under the greatest pressure. Fundamentally, a farmer is paid the difference between the value of their land for farming versus development. In return, the landowner agrees to a conservation easement which restricts the use of the property to agricultural purposes. The program is financed through a 1.5 per cent property tax to generate $ 3.5 million dollars a year.

The Big Darby Creek project in Ohio also strives to protect a largely intact aquatic system which supports a large concentration of rare species. About 80 per cent of the upland area is dedicated to maize and soybean production. Like Virginia Beach, intense development pressure from the city of Columbus is posing a major threat to the river system. Concern for the river has sparked a phenomenal public and private collaborative effort to protect the water quality. In addition, many local farmers have embraced no-till agriculture as a means of reducing sedimentation. The United States Department of Agriculture (USDA) has also played an instrumental role by focusing financial and technical assistance in the watershed. Since 1991, the project has attracted over $ 14 million dollars in grants and loans to solve resource issues.

The third case study focuses on the State of Nebraska where two different policy approaches are employed. In the Sandhill area of North-central Nebraska, TNC is grazing bison to restore the natural processes which support a native prairie ecosystem. Profitable niche markets for bison meat, in addition to cattle grazing have permitted TNC to make the preserve financially self-supporting. In Southern Nebraska, along the Platte River, TNC is working in co-operation with the State of Nebraska, the University of Nebraska and ConAgra, a private multi-national agribusiness, to demonstrate farming practices which can improve the riparian corridor, maintain profitability and support habitat for migratory sandhill crane.

In the final case study, TNC is assisting ranchers in a one million acre area along the border of Arizona and New Mexico to restore a healthy grassland ecosystem with the associated benefit of improving ranching profitability. The landowners are under intense pressure to sell their ranches for second home development. Maintaining an intact landscape and ranching is an important goal to TNC and the community. Land ownership in the area is split between private, state and federal entities which has created problems for co-ordinated management in the past. One of the most important activities associated with this project is the technical assistance provided by USDA and the willing co-operation of the public land managers.

TNC's private sector initiatives are greatly enhanced by the work of public agencies and the targeting of government programs. The public agencies provide technical expertise that is often expensive or unavailable to locally grown conservation initiatives. A key factor of success in many TNC project areas is willing co-operation among the public agencies. Avoiding jurisdictional battles and permitting flexible program implementation that accommodates local conditions is extremely important. Equally important are the public funds that are used as incentives to encourage environmentally sound management practices, underwrite farmland protection and provide money for planning and management. Government programs that fund innovation and technology research and provide grants for new product development and market identification are critical to identifying economically feasible and environmentally sustainable technologies and products.

16.   United States: Wetlands conservation

The United States' experience in wetlands conservation holds valuable lessons for any nation seeking a productive and environmentally sound agriculture. Those lessons include finding the right mix of voluntary and regulatory approaches to conservation; bringing together programs and authorities that were established piecemeal and making a cohesive set of policies and programs that is affordable and acceptable to society; co-ordinating the work of different federal agencies; focusing on natural systems, not just one resource at a time; and coping, as a society, with a contentious environmental issue, balancing public and private interests to ensure the long-term sustainability of our natural resources.

During the broad-based environmental movement of the 1960's and 1970's, American agriculture and society at large became sensitised to the importance of wetlands as natural filters, natural flood control, sources of soil moisture in times of drought, harbours of biological diversity, and beautiful landscapes as well as alternate sources of income. It was discovered that after more than a century of public policy that promoted drainage, the continental United States had only about half of the wetlands that had existed when the first European settlers arrived. Thus began an intense public effort to reduce wetland destruction, and between the mid-1970's and 1992, the average rate of wetland conversion for agricultural production plummeted from about 398 000 acres per year to about 31 000 acres per year.

The progress in U.S. wetlands conservation is largely due to the grass-roots effort of the voluntary soil-and-water conservation movement initiated by private landowners and the U.S. Department of Agriculture (USDA) in the 1930's. Increasing regulation of wetlands and new incentive and disincentive programs have made a significant impact also. Particularly effective are: (i) the National Environmental Policy Act of 1969, a federal law that requires an assessment of the environmental consequences of federal policies and programs; (ii) the Federal Water Pollution Control Act (later retitled as the Clean Water Act), which requires farmers who want to drain wetlands or discharge dredge or fill material into U.S. waters (including wetlands) to get a permit from the U.S. Army Corps of Engineers; (iii) the Water Bank Program, which was offered in 10 states that have major migratory waterfowl flyways; (iv) the 1985 Food Security Act, which tied eligibility for most federal farm program benefits (including commodity price supports, agricultural credit, and crop insurance) to the application of land stewardship practices, including the protection of wetlands; (v) the Tax Reform Act of 1986, which affected capital gains and reduced tax advantages from land conversion; (vi) the Food, Agriculture, Conservation, and Trade Act of 1990, which strengthened the Food Security Act but made it more flexible and established a Wetlands Reserve Program under which farmers could voluntarily sell land easements to the government for wetlands restoration purposes; (vii) the Federal Agriculture Improvement and Reform Act of 1996, which consolidated several conservation programs that had evolved piecemeal, thus increasing efficiency and ensuring that public funds serve local people and the resource concerns of highest priority.

Increasing flexibility built into wetland conservation provisions of farm legislation since 1985 responded to landowner frustration over several contentious issues, including rigidity of the law regarding loss of farm program benefits with a single violation; overlapping federal responsibilities and conflicting standards; private property rights; and landowner frustration with basic culture changes in their relationship with USDA - from a purely voluntary-program relationship to a more regulatory one.

The U.S. has a reasonably accurate picture of broad national trends in wetlands condition. But success monitoring is a new area for science, and we seek a better understanding of how wetlands work and just what it is that we are trying to assess. This means expanding knowledge on wetland functions, wetland values, and wetland health. To that end, Natural Resources Conservation Service is working with other federal agencies and the private sector to develop regionally based techniques for wetland functional assessment using a hydrogeomorphic approach. Computer modelling and geographic information systems are important tools in the move to decision support systems and to monitoring real-time events and measuring outcomes of our programs, policies, and conservation practices at the local and national levels.

It's expected that the U.S. soon will experience a net increase in wetlands on the agricultural landscape as a result of the lessons learned: (i) base value decisions and policies on good science; (ii) co-ordinate resource-protection programs and resource data gathering across agency lines, and target program resources to the areas where they are most needed; (iii) use regulatory programs sparingly and carefully; (iv) promote conservation policies and programs that are voluntary and incentive-based and that stress partnerships and local involvement; (v) involve all stakeholders at the community and watershed levels; (vi) understand that no nation will have lasting conservation on private lands until landowners are excited about the land and understand that environmentally sound land use is not a limit on personal freedom but rather a positive exercise of skill and insight.

17.   Australia: Policy approach for managing environmental benefits and costs of a sustainable agriculture

The general approach of Australian governments towards environmental benefits and costs of sustainable agriculture is guided by a number of general principles. These principles include a recognition that property ownership bestows both rights and obligations upon the property owner. In light of this recognition, both incentives and disincentives, as well as the provision and dissemination of information, may be utilised in policy measures to ensure that agri-environmental outcomes (i.e. the supply of agri-environmental attributes) meet adequacy standards established by the community. Therefore, policy measures should directly address the causes, rather than the symptoms, of any perceived inadequacy in the level or quality of an agri-environmental attribute, i.e. measures should address the underlying market failure (or policy failure) that is causing the inadequacy of supply.

Policy measures should not be linked to the volume or type of agricultural production, i.e. measures should be as minimally production (and hence trade) distorting as possible. This principle is based on concerns to see that measures are cost effective, do not contravene international obligations (especially World Trade Organisation obligations) and have minimal adverse effects on the economies of agriculture dominated developing countries. Policy measures should be made available to all landowners, i.e. they should not be restricted to farmers. Measures should also be transparent and have clear objectives, i.e. the desired quantity and quality of the attribute being promoted should be defined precisely. Policy instruments should contain measurable performance indicators which are regularly evaluated for effectiveness; and they should be the most cost effective means of achieving the desired level of attribute provision.

Many agri-environment problems are the result of insufficient awareness on the part of landowners or a lack of practical means to address problems. In such cases, it is worth exploring the potential for measures such as research and development activities and voluntary co-operative approaches (including education and awareness raising activities) which may address the underlying cause of the inadequacy of supply of an agri-environmental attribute. In cases where there is a conflict of views between landowners and the wider community, as to whether an agri-environmental attribute is undersupplied or oversupplied, an alternative approach is the provision of direct income payments to landowners who provide (or agree to maintain) certain desired levels of particular agri-environmental attributes. It is noted that it is virtually impossible to provide direct income payments without distorting production decisions, hence there is a need for such measures to be carefully designed and tightly targeted. In cases were landowners fail to meet the adequacy standards established by the community, disincentive measures, such as regulations and taxes, may represent the appropriate policy response. In these circumstances, the Polluter Pays Principle, with its focus on internalising negative externalities, provides a particularly useful approach for policy makers.

In Australia, decisions about the use of agricultural resources are made largely by private individuals acting on the basis of information and incentives provided through markets within an established institutional framework. This institutional framework includes a strong, but not absolute, recognition of (and entitlement to) landowner "rights". These "rights" can and have at various times been restricted by Australian governments at Commonwealth (federal), state/territory and local levels through application of their direct and delegated powers. Australian policies take into account the fact that private (landowner) and community interests do not always coincide and that there is a role for policy makers to address the market failures that underlie such divergences. These market failures largely take the form of underprovision of public goods or the presence of negative externalities arising from agricultural production activities.

In the main, the Australian approach seeks to capitalise on the widespread coincidence of private (landowner) valuation of agri-environmental attributes and social (community) valuation of agri-environmental attributes. Policies seek to address the reasons why landowners are inadequately supplying (or in the case of off-farm attributes, contributing to the inadequacy of supply of) attributes that they themselves desire more of. Policy measures are directed towards establishing effective pricing and property rights systems that encourage resource users to take into account all economic and environmental costs and benefits and which, therefore, encourage the best long-term use of natural resources. In instances where markets do not operate effectively, regulatory systems necessary for the enhancement of environmental quality are established. Policy also seeks to provide information and promote the skills that resource owners and managers need to make appropriate decisions, and to support research and development to improve the information base, the technology and the management practices relevant to natural resource management; as well as to support activities such as the preparation of catchment and regional plans that are required to address large scale resource management concerns.

The overall policy framework for the approach of Australian Governments has a number of elements: (i) the Decade of Landcare Plan, agreed to by the Commonwealth and all State and Territory governments, which is a long-term national policy designed to bring about environmentally sustainable and economically viable land management; (an example of a voluntary co-operative approach); (ii) a policy framework agreed to by the Commonwealth and all State and Territory Governments through the Council of Australian Governments (COAG) which provides for a national approach to institutional and microeconomic reforms to encourage ecologically sustainable and economically viable management of water resources; (iii) consistent with the Decade of Landcare

Plan and the COAG water management policy, the National Landcare Program, administered by the Commonwealth Government, provides financial support to the States and Territories and community groups for natural resource management and environmental conservation activities that primarily provide public benefits; (iv) the Murray-Darling Basin Initiative (MDBI) which provides for a joint approach by Commonwealth and State governments for natural resource management in the Murray-Darling Basin, which covers one-seventh of Australia's land mass, encompasses several States and produces a significant proportion of Australia's agricultural production.

18.   New Zealand: The environmental effects of removing agricultural subsidies

In many cases environmental problems associated with agricultural production can be attributed to conflicting government policies. Government programmes to address environmental problems caused by agricultural production can be undermined by government policies which encourage the intensification of agriculture. The approach of the New Zealand Government towards agriculture and its effects on the environment has been to create a balanced and coherent policy environment which provides for sound environmental management undistorted by government funded agricultural support programmes.

In the past decade, New Zealand has implemented wide-ranging economic and environmental reforms. Government intervention in the economy, including in agriculture, had led to severe misallocation of resources and high levels of assistance which could no longer be maintained. In 1984 and succeeding years, government assistance to agriculture was virtually eliminated. Following the removal of subsidies, livestock numbers declined, the use of fertilizers and pesticides decreased, and there was an increase in afforestation as increasing returns to forestry were reflected at the farm level. Conversion of regenerating and established native forest to agriculture has virtually ceased, as has development of new irrigation and drainage schemes. All of these changes lessen the likelihood of farming systems causing degradation of marginal lands and off-site contamination of water resources.

Several other reforms have had implications for agriculture and the environment. For example, the Government has largely withdrawn from providing disaster relief to farmers, and now has a policy of encouraging individual landholders to manage climatic risks. Expenditure on disaster relief for agriculture has declined from an annual average of NZ$ 26 million in the five years ending 1990/91 to NZ$ 5.6 million in 1992/93 and none in 1994/95 or 1995/96.   More recently, the Biosecurity Act 1993 was enacted to provide for pest management, including mechanisms to secure funding from beneficiaries and those who exacerbate a pest problem. Some funding continues from both central and regional government, but it is expected that the share of non-government funding will increase.

Various environmental issues remain to be addressed. The experience of agricultural reform has confirmed for New Zealand that the removal of agricultural subsidies is necessary but not sufficient to address the environmental impacts of agriculture. Targeted domestic environmental policies are necessary to address the environmental effects of agriculture. User-pays and polluter-pays principles are well established in New Zealand. For a range of activities, farmers must obtain environmental permits and, in an increasing number of circumstances, must pay the administrative costs and on-going monitoring costs associated with a permit. Apart from a special programme to address severe land degradation in one region of the country, and a diminishing number of regional council cost-sharing programmes for soil conservation, there are no government subsidies to farmers to improve environmental performance.

New Zealand has taken the approach that it is necessary to remove distorting price signals and address environmental "bads" before considering whether agriculture provides environmental "goods" that require government assistance. To have done otherwise, i.e. to compensate farmers for perceived environmental "goods" without addressing the "bads", would have risked entrenching current systems that were causing environmental damage, and would have been just another way to subsidise farming. There is little evidence of market failure in the provision of environmental "goods" by New Zealand agriculture. In New Zealand, most biodiversity resides in natural ecosystems, both terrestrial and aquatic. While agriculture can provide landscape amenities and *in-situ* preservation of biodiversity, including biodiversity of agricultural species, these "goods" are by-products of agricultural systems and are still being provided despite the withdrawal of government assistance. If anything, government assistance to agriculture was having a negative effect on the supply of biodiversity and other environmental goods, and the first step was to remove such distortions. Although there is still some way to go, New Zealand is moving towards internalisation of environmental costs in order to encourage the efficient and sustainable use of natural resources.

19. New Zealand: Policy considerations regarding landscape amenities and biodiversity from sustainable agriculture

Many features of farms which provide landscape amenities and biological diversity ("biodiversity") also provide goods or services to farmers. Small parcels of forest provide firewood and hunting opportunities. Forests and hedgerows provide shelter for livestock; riparian vegetation enhances fish habitat. Features which enhance the visual appeal of the landscape make the farm itself more attractive, enhancing the value of the property. Beyond farm boundaries, however, agriculture has primarily a negative effect on biodiversity, due to farm runoff of sediment, fertiliser and pesticides.

Economic theory suggests that public goods may be undersupplied by market forces. Although landscape amenities and biodiversity have attributes of public goods, in some cases sufficient quantities may be provided as by-products of agricultural production, and there would be no market failure. Thus, before intervening, policy makers must first decide whether there is in fact a sub-optimal supply of landscape and biodiversity, based on the desired quality and quantity of these goods and the costs and benefits of their provision.

Another question is how the supply of these by-products would change under agricultural policy reform. It seems likely that output-linked subsidies have an adverse effect on landscape, because they provide an enhanced financial incentive to convert hedgerows, stream banks, wetlands etc. into productive area. Such subsidies can also encourage emphasis on one or only a few crops, as well as mechanisation and increased use of inputs, to the detriment of biodiversity. Hence, the removal of output-linked subsidies should lessen the pressure on these resources and enhance the diversity of crops and other farm enterprises. Other considerations are also noteworthy. For instance, a decision to compensate farmers for redefining property rights will tend to reinforce the myth that rights are absolute, making it even more difficult to achieve environmental objectives in the future. Paying farmers for what they arguably should do (and in most cases are already doing) undermines the duty of stewardship for which many feel landholders should be responsible.

Policies to secure environmental benefits from agriculture should be preceded by, or at least be implemented together with, policies to address environmental harm caused by agriculture. This promotes a sustainable agriculture, and ensures that policy does not encourage current systems that are causing environmental damage. Any payments for environmental goods should be offset by penalties for environmental bads, i.e. payments should be for "net" environmental services. Governments should set reference levels for environmental goods subject to intervention; landowners should be required to achieve the reference level, and payments should only be available for goods supplied in excess of that level.

Interventions should seek to alter supply or demand conditions permanently so that the desired amenity will continue to be enjoyed at minimum or even zero economic cost in future. Appropriate measures might include regulation, information and facilitation, land purchases or permanent easements, mechanisms that allow providers of amenities to charge beneficiaries, or direct payments to producers. Direct payments should only be used where necessary to achieve the programme's environmental objective, taking account of available alternatives. To the maximum extent possible, in order to avoid production and trade distortions, payments should not be related to, or based on, the type or volume of production, or any factors of production in a manner which would affect the type or volume of production. Rather, payments should be based on environmental outcomes, or on farming practices which are primary determinants of those outcomes. Payments should be available to anyone who can provide the environmental good in question, and should not be restricted to farmers. Payments should be limited to the extra costs or net loss of income involved in complying with the government programme.

New Zealand uses a variety of approaches to address agriculture-related biodiversity and landscape issues. Instruments used include research and facilitation, regulation (polluter-pays), user charges for services, permanent easements and selective purchase. In addition, there are a variety of voluntary, co-operative initiatives by the private sector. This approach enables the New Zealand government to promote environmental benefits from a sustainable agriculture, without distorting production decisions and without reinforcing notions of absolute property rights that would complicate environmental policy in the longer term.

20. Japan: The environmental benefits from agriculture in the Asian monsoon climate zone and policy implications for their maintenance and enhancement

Countries within the Asian monsoon climate zone, which stretch from east Asia to south-east Asia, have several common characteristics in terms of natural environment, such as a seasonal climate variation with high humidity, high precipitation, and steep topography. For example, in Japan, precipitation is almost twice that of other OECD countries and is concentrated in the very short summer period. Therefore, it is obvious that, if the land in the Asian climate zone is left without appropriate management practices, water resources and surface soil would rapidly disappear toward the sea without being used effectively.

In the Asian monsoon climate zone, people have developed the wetland rice production system, which is adapted to the climatic conditions. The rice paddy system with flooding cultivation in the growing season has several features, such as continuous cropping without any significant environmental problem and maintenance of soil fertility. Being able to support a greater population than other farming systems, this system has, for a long time, formed the basis of rural communities in the Asian monsoon climate zones. The main beneficial effects on the environment of the rice paddy fields are, flood prevention, fostering of water resources, landslide prevention, soil erosion prevention, air purification, maintenance of biodiversity, and formation of scenic landscape and amenities. Moreover, some of these benefits, especially the prevention of floods and land subsidence, also contribute to the environmental protection of the lowland urban areas. Therefore, abandoning rice paddy fields would lead to a significant reduction in environmental benefits from agriculture.

Paddy fields and agricultural production facilities can store a considerable amount of rain water, which can prevent flood damage despite intensive rainfall. Rain and irrigation water supplied to agricultural land penetrate into the soil and contribute to fostering of water resources. Terraced paddy fields, managed by farmers' daily work, prevent landslides in mountainous regions, which occupy over 70 per cent of total land. Well-managed paddy fields suffer very little from soil erosion because irrigation surface water can absorb the impact of rainfall. In addition, paddy field areas provide both inhabitants and visitors with traditional scenic landscapes formed through farmers' daily activities, as well as opportunities for recreation, relaxation and reflection. They also form and preserve habitable areas within a calm and pleasant environment, and preserve the unique culture and tradition of their long history.

Moreover, the adverse effects on the environment of agricultural inputs, such as fertilizers and pesticides, are markedly low. In paddy soil, ammonium nitrogen, applied as nitrogen fertilizer, tends to be absorbed into soil particles and hardly leached into the ground. In addition, rice paddy fields have the function of absorbing the run-off from animal husbandry, fruit, vegetables and other upland crops which are located at higher altitude. Therefore, surveys revealed that the nitric nitrogen content of underground water near paddy fields was remarkably low in spite of considerable nitrogen fertilizer application.

Because of the environmental benefits associated with agricultural activities in the Asian monsoon climate zone, policies to maintain appropriate agricultural production and land management are required. It is recognised that the reduction in agricultural protection through trade liberalisation has caused a depression in agriculture, which has led to the abandoning of farm land and the loss of environmental benefits. In particular, it is considered that in conserving environmental benefits, the following points have to be taken into account: (i) most of the environmental benefits of agriculture in Japan are lost when sound agricultural practices are relaxed or abandoned, appropriate farming practices to maintain environmental benefits, are required; (ii) it would be very difficult or would require huge public expenditure to maintain the environmental benefits of agriculture in Japan by alternative land use, such as forestry, and physical infrastructures; (iii) the introduction of direct payments for the environment may be difficult, as there are few reliable methods for quantifying the environmental benefits from agriculture in monetary terms, and their implementation would entail costs for administration, and for estimation of the payment to each of a great number of small farms.

21.   Korea: Sustainable soil management by farm parcel

Korean soils are mostly sandy, low in organic matter content and fertility, and susceptible to erosion, because more than 60 per cent of soils are located on sloping areas and are exposed to violent rainstorms in the summer. The government has implemented several long-term projects of soil surveys and soil fertility, with food self-sufficiency in mind, resulting in a considerable improvement of soil quality. A soil survey was carried out on every parcel of paddy field, covering 1.4 million hectares, and soil testing was done on 617 000 samples from 6.7 million farm parcels. Information for soil fertilization were issued to every farmer, based on the soil survey and chemical analysis from 1980 to 1989. A five year upland programme has been conducted on 583 000 hectares since 1995. In this way, a basis for good soil management by farm parcel has been established.

Present day agriculture, due to the high inputs of chemical and organic fertilizers, has made it possible to increase production; however it has also led to the pronounced increase of available phosphate and potassium in soils. Salt accumulation is very high, especially in greenhouse soils. Sustainable soil management by every farm is not only essential for environment conservation, but also for productivity maintenance. There are 144 rural county extension offices are operating soil testing laboratories for soil samples from farm parcels. Each farmer is given the information on recommendation on fertilizers, with reports on soil quality, salt accumulation, amelioration techniques, and the required quantity of fertilizers needed. Soil management based on the soil analysis could improve and correct various agricultural problems, including low productivity, salt accumulation, and even heavy metal pollution of farm parcels. To minimise environmental impacts of the recent high inputs of agriculture, optimal fertilization by each farm parcel is essential for sustainable soil management.

LIST OF PARTICIPANTS/LISTE DES PARTICIPANTS

HELSINKI, 10-13 SEPTEMBER/SEPTEMBRE 1996

**AUSTRALIA/
AUSTRALIE**

Volker AEUCKENS, Department of Primary Industries and Energy, Canberra
David PURCELL, Delegation of Australia to the OECD, Paris

**AUSTRIA/
AUTRICHE**

Helmut WALTER, Federal Ministry of Agriculture, Vienna
Walter KUCERA, Austrian Chambers of Agriculture, Vienna

**BELGIUM/
BELGIQUE**

Jacques DETROZ, Ministère de l'agriculture, Bruxelles
Joseph DEMUYNCK, Ministry of the Flemish Community Environment
  Management Administration, Brussels
Jean-Pierre LAPERCHES, General Directorate for Environment and Natural
  Resources
Hilde VANDENDRIESSCHE, Soil Service of Belgium, Heverlee

CANADA

Ian CAMPBELL, Agriculture Canada, Ontario

**FINLAND/
FINLANDE**

John Holger SUMELIUS, Agricultural Economics Research Institute, Helsinki
Heikki LATOSTENMAA, Ministry of Environment, Helsinki
Tiina MALM, Ministry of Agriculture, Helsinki
Satu NURMI, Ministry of Environment, Helsinki
Veili-Pekka TALVELA, Ministry of Agriculture, Helsinki
Risto TIMONEN, Ministry of Agriculture, Helsinki
Markku TORNBERG, Central Union of Agricultural Producers and Forest Owners,
  Helsinki
Hans VOGT, The Finnish Association for Nature Conservation, Helsinki
Jyri AAKKULA, Agricultural Economics Research Institute, Helsinki
Jyri OLLILA, Delegation of Finland to the OECD, Paris
Seija HAKKARAINEN, Ministry of Agriculture and Forestry, Helsinki
Harri WESTERMARCK, University of Helsinki

FRANCE

Dominique GAGEY, Ministère de l'agriculture, de la pêche et de l'alimentation,
  Paris
Guy POIRIER, Ministère de l'agriculture, de la pêche et de l'alimentation, Paris
Joseph RACAPÉ, Ministère de l'environnement, Paris
Pierre RAINELLI, Institut National de la Recherche Agronomique, Paris

**GERMANY/
ALLEMAGNE**

Hartmunt HÖH, Federal Statistical Office, Wiesbaden
OSTERMEYER-SCHLÖDER, Federal Ministry of Environment, Bonn
Reiner PLANKL, Federal Agricultural Research Centre, Braunschweig
Petra STEFFENS, Federal Ministry of Food, Agriculture and Forestry, Bonn

GREECE/GRÈCE

Marlena TIKOF, Ministry of Agriculture, Athens

HUNGARY/ **HONGRIE**	Csaba NEMES, Ministry of Environment and Regional Policy, Budapest József SURJÁN, Ministry of Agriculture, Budapest
IRELAND/ **IRLANDE**	Dan GAHAN, Department of Agriculture, Food and Forestry, Dublin
ITALY/ITALIE	Angelo INNAMORATI, Ministère des Ressources Agricoles, Rome
JAPAN/JAPON	Ryohei KADA, Kyoto University, Kyoto Michinori NISHIO, Ministry of Agriculture, Forestry and Fisheries, Tsukuba Masamichi SAIGO, Ministry of Agriculture, Forestry and Fisheries, Tokyo Shuji YAMADA, Permanent Delegation of Japan to the OECD, Paris Yasuo WATANABE, Ministry of Agriculture, Forestry and Fisheries, Tokyo

NETHERLANDS/
PAYS-BAS
	Gérard van DIJK, Ministry of Agriculture, Fisheries and Nature Management,
	   The Hague
	Johan HEINEN, Ministry of Agriculture, Fisheries and Nature Management, Utrecht
	Jan van der KOLK, Ministry of Housing, Spatial Planning and Environment,
	   The Hague
	Jacob NIEUWENHUIZE, Ministry of Agriculture, Fisheries and Nature
	   Management, The Hague
	G. WESTENBRIK, Ministry of Agriculture, Fisheries and Nature Management,
	   The Hague

NEW ZEALAND/ **NOUVELLE-** **ZÉLANDE**	Mr. Jim SINNER, Ministry of Agriculture, Wellington
NORWAY/ **NORVÈGE**	Gabriella DÅNMARK, Ministry of Agriculture, Oslo Morten HAUGERUD, Ministry of Agriculture, Oslo Frode LYSSANDRAE, Ministry of Agriculture, Oslo
PORTUGAL	N. GALHARDO, Ministry of Agriculture, Lisbon
SWEDEN/SUÈDE	Rolf AKESSON, Ministry of Agriculture, Stockholm Goran BOBERG, Ministry of Agriculture, Stockholm Peter EINARSON, The Swedish Association for Nature Conservation, Stockholm Pernilla IVARSSON, Ministry of Agriculture, Stockholm Karl Johan LIDEN, Swedish Board of Agriculture, Jonkoping
SWITZERLAND/ **SUISSE**	Brigitte DECRAUSAZ, Office fédéral de l'agriculture, Bern Ariane SOTOUDEH, Office fédéral d'environnement, des forêts et du paysage, Berne Heidi BRAVO, Union Suisse des Paysans, Brugg
UNITED **KINGDOM/** **ROYAUME-UNI**	Heather BLAKE, Ministry of Agriculture, Fisheries and Food, London Robert DAVIES, Department of the Environment, London J.P. MURIEL, Ministry of Agriculture, Fisheries and Food, London Michael HARRISON, Ministry of Agriculture, Fisheries and Food, London Stephen REEVES, Department of the Environment, London

UNITED STATES/	Paul JOHNSON, U.S. Department of Agriculture, Washington
ÉTATS-UNIS	Peter F. SMITH, U.S. Department of Agriculture, Washington
	John STIERNA, U.S. Department of Agriculture, Washington
	Diane VOSICK, The Nature Conservancy, Arizona

COMMISSION OF THE EUROPEAN COMMUNITIES/
COMMISSION DES COMMUNAUTÉS EUROPÉENES

Hans Christian BEAUMOND, Directorate General VI/Agriculture, Brussels
Martin SCHEELE, Directorate General VI/Agriculture, Brussels
Eric BIGNAL, Directorate General VI/Agriculture, Brussels
Reinhard PRIEBE, Directorate General VI/Agriculture, Brussels
Frank FAY, Directorate General VI/Agriculture, Brussels

OBSERVERS/OBSERVATEURS

KOREA/	Doo Bong HAN, Korea University, Seoul
CORÉE	Song-Soo LIM, Korea Rural Economic Institute, Seoul
	Moo-Eon PARK, Ministry of Agriculture and Forestry, Suwon
	Jae-Sung SHIN, National Institute of Agricultural Science and Technology, Seoul

NON-GOVERNMENTAL ORGANISATIONS/ORGANISATIONS NON GOUVERNEMENTALES

European Confederation of Agriculture/Confédération européene pour l'agriculture
Henri SUTER, Union Suisse des Paysans, Brugg

Institute for European Environment Policy/Institut pour une politique européene de l'environnement
David BALDOCK, London

International Federation of Agricultural Producers/Fédération internationale des producteurs agricoles
Frances KINNON, France

The Royal Society for the Protection of Birds/Société Royale pour la protection des oiseaux
Miguel A. NAVESO, BirdLife Spain, Madrid
Vicki SWALES, Birdlife International, Bedfordshire

World Wide Fund for Nature/Fonds mondial pour la nature
Natacha YELLACHICH, WWF European Policy Office, Brussels
Hilmar von MUENCHHAUSEN, WWF Germany, Frankfurt am Main

CONSULTANTS

Daniel W. BROMLEY, University of Wisconsin, Madison
Sandra S. BATIE, Michigan State University, East Lansing
Peter Lewis NOWICKI, European Centre for Nature Conservation, Tilburg
George HUTCHINSON, Queen's University, Belfast
Anton D. MEISTER, Massey University, Palmerston North

OECD SECRETARIAT/SECRÉTARIAT DE L'OCDE
Directorate for Food, Agriculture and Fisheries/Direction de l'alimentation, de l'agriculture et des pêcheries
Gérard VIATTE, Director
Wilfrid LEGG, Head of Division
Luis PORTUGAL, Principal Administrator
Jane KYNASTON, Secretary
Françoise BÉNICOURT, Secretary

Environment Directorate/Direction de l'environnement
Michel POTIER, Head of Division

Territorial Development Service/Service du développement territorial
Christian HUILLET, Principal Administrator

FINNISH ORGANISERS/ORGANISATEURS FINLANDAIS
Agricultural Economics Research Institute
John SUMELIUS, Senior Research Co-ordinator
Meri JUNTTI, Secretary

MAIN SALES OUTLETS OF OECD PUBLICATIONS
PRINCIPAUX POINTS DE VENTE DES PUBLICATIONS DE L'OCDE

AUSTRALIA – AUSTRALIE
D.A. Information Services
648 Whitehorse Road, P.O.B 163
Mitcham, Victoria 3132   Tel. (03) 9210.7777
                          Fax: (03) 9210.7788

AUSTRIA – AUTRICHE
Gerold & Co.
Graben 31
Wien I   Tel. (0222) 533.50.14
                          Fax: (0222) 512.47.31.29

BELGIUM – BELGIQUE
Jean De Lannoy
Avenue du Roi, Koningslaan 202
B-1060 Bruxelles   Tel. (02) 538.51.69/538.08.41
                          Fax: (02) 538.08.41

CANADA
Renouf Publishing Company Ltd.
5369 Canotek Road
Unit 1
Ottawa, Ont. K1J 9J3   Tel. (613) 745.2665
                          Fax: (613) 745.7660
Stores:
71 1/2 Sparks Street
Ottawa, Ont. K1P 5R1   Tel. (613) 238.8985
                          Fax: (613) 238.6041

12 Adelaide Street West
Toronto, QN M5H 1L6   Tel. (416) 363.3171
                          Fax: (416) 363.5963

Les Éditions La Liberté Inc.
3020 Chemin Sainte-Foy
Sainte-Foy, PQ G1X 3V6   Tel. (418) 658.3763
                          Fax: (418) 658.3763

Federal Publications Inc.
165 University Avenue, Suite 701
Toronto, ON M5H 3B8   Tel. (416) 860.1611
                          Fax: (416) 860.1608

Les Publications Fédérales
1185 Université
Montréal, QC H3B 3A7   Tel. (514) 954.1633
                          Fax: (514) 954.1635

CHINA – CHINE
Book Dept., China National Publications
Import and Export Corporation (CNPIEC)
16 Gongti E. Road, Chaoyang District
Beijing 100020   Tel. (10) 6506-6688 Ext. 8402
                          (10) 6506-3101

CHINESE TAIPEI – TAIPEI CHINOIS
Good Faith Worldwide Int'l. Co. Ltd.
9th Floor, No. 118, Sec. 2
Chung Hsiao E. Road
Taipei   Tel. (02) 391.7396/391.7397
                          Fax: (02) 394.9176

**CZECH REPUBLIC –
RÉPUBLIQUE TCHÈQUE**
National Information Centre
NIS – prodejna
Konviktská 5
Praha 1 – 113 57   Tel. (02) 24.23.09.07
                          Fax: (02) 24.22.94.33
E-mail: nkposp@dec.niz.cz
Internet: http://www.nis.cz

DENMARK – DANEMARK
Munksgaard Book and Subscription Service
35, Nørre Søgade, P.O. Box 2148
DK-1016 København K   Tel. (33) 12.85.70
                          Fax: (33) 12.93.87

J. H. Schultz Information A/S,
Herstedvang 12,
DK – 2620 Albertslung   Tel. 43 63 23 00
                          Fax: 43 63 19 69
Internet: s-info@inet.uni-c.dk

EGYPT – ÉGYPTE
The Middle East Observer
41 Sherif Street
Cairo   Tel. (2) 392.6919
                          Fax: (2) 360.6804

FINLAND – FINLANDE
Akateeminen Kirjakauppa
Keskuskatu 1, P.O. Box 128
00100 Helsinki

Subscription Services/Agence d'abonnements :
P.O. Box 23
00100 Helsinki   Tel. (358) 9.121.4403
                          Fax: (358) 9.121.4450

*FRANCE
OECD/OCDE
Mail Orders/Commandes par correspondance :
2, rue André-Pascal
75775 Paris Cedex 16   Tel. 33 (0)1.45.24.82.00
                     Fax: 33 (0)1.49.10.42.76
                     Telex: 640048 OCDE
Internet: Compte.PUBSINQ@oecd.org

Orders via Minitel, France only/
Commandes par Minitel, France exclusivement :
36 15 OCDE

OECD Bookshop/Librairie de l'OCDE :
33, rue Octave-Feuillet
75016 Paris   Tel. 33 (0)1.45.24.81.81
                          33 (0)1.45.24.81.67
Dawson
B.P. 40
91121 Palaiseau Cedex   Tel. 01.89.10.47.00
                          Fax: 01.64.54.83.26

Documentation Française
29, quai Voltaire
75007 Paris   Tel. 01.40.15.70.00

Economica
49, rue Héricart
75015 Paris   Tel. 01.45.78.12.92
                          Fax: 01.45.75.05.67

Gibert Jeune (Droit-Économie)
6, place Saint-Michel
75006 Paris   Tel. 01.43.25.91.19

Librairie du Commerce International
10, avenue d'Iéna
75016 Paris   Tel. 01.40.73.34.60

Librairie Dunod
Université Paris-Dauphine
Place du Maréchal-de-Lattre-de-Tassigny
75016 Paris   Tel. 01.44.05.40.13

Librairie Lavoisier
11, rue Lavoisier
75008 Paris   Tel. 01.42.65.39.95

Librairie des Sciences Politiques
30, rue Saint-Guillaume
75007 Paris   Tel. 01.45.48.36.02

P.U.F.
49, boulevard Saint-Michel
75005 Paris   Tel. 01.43.25.83.40

Librairie de l'Université
12a, rue Nazareth
13100 Aix-en-Provence   Tel. 04.42.26.18.08

Documentation Française
165, rue Garibaldi
69003 Lyon   Tel. 04.78.63.32.23

Librairie Decitre
29, place Bellecour
69002 Lyon   Tel. 04.72.40.54.54

Librairie Sauramps
Le Triangle
34967 Montpellier Cedex 2   Tel. 04.67.58.85.15
                            Fax: 04.67.58.27.36

A la Sorbonne Actual
23, rue de l'Hôtel-des-Postes
06000 Nice   Tel. 04.93.13.77.75
                          Fax: 04.93.80.75.69

GERMANY – ALLEMAGNE
OECD Bonn Centre
August-Bebel-Allee 6
D-53175 Bonn   Tel. (0228) 959.120
                          Fax: (0228) 959.12.17

GREECE – GRÈCE
Librairie Kauffmann
Stadiou 28
10564 Athens   Tel. (01) 32.55.321
                          Fax: (01) 32.30.320

HONG-KONG
Swindon Book Co. Ltd.
Astoria Bldg. 3F
34 Ashley Road, Tsimshatsui
Kowloon, Hong Kong   Tel. 2376.2062
                          Fax: 2376.0685

HUNGARY – HONGRIE
Euro Info Service
Margitsziget, Európa Ház
1138 Budapest   Tel. (1) 111.60.61
                          Fax: (1) 302.50.35
E-mail: euroinfo@mail.matav.hu
Internet: http://www.euroinfo.hu//index.html

ICELAND – ISLANDE
Mál og Menning
Laugavegi 18, Pósthólf 392
121 Reykjavik   Tel. (1) 552.4240
                          Fax: (1) 562.3523

INDIA – INDE
Oxford Book and Stationery Co.
Scindia House
New Delhi 110001   Tel. (11) 331.5896/5308
                          Fax: (11) 332.2639
E-mail: oxford.publ@axcess.net.in

17 Park Street
Calcutta 700016   Tel. 240832

INDONESIA – INDONÉSIE
Pdii-Lipi
P.O. Box 4298
Jakarta 12042   Tel. (21) 573.34.67
                          Fax: (21) 573.34.67

IRELAND – IRLANDE
Government Supplies Agency
Publications Section
4/5 Harcourt Road
Dublin 2   Tel. 661.31.11
                          Fax: 475.27.60

ISRAEL – ISRAËL
Praedicta
5 Shatner Street
P.O. Box 34030
Jerusalem 91430   Tel. (2) 652.84.90/1/2
                          Fax: (2) 652.84.93

R.O.Y. International
P.O. Box 13056
Tel Aviv 61130   Tel. (3) 546 1423
                          Fax: (3) 546 1442
E-mail: royil@netvision.net.il

Palestinian Authority/Middle East:
INDEX Information Services
P.O.B. 19502
Jerusalem   Tel. (2) 627.16.34
                          Fax: (2) 627.12.19

ITALY – ITALIE
Libreria Commissionaria Sansoni
Via Duca di Calabria, 1/1
50125 Firenze   Tel. (055) 64.54.15
                          Fax: (055) 64.12.57
E-mail: licosa@ftbcc.it

Via Bartolini 29
20155 Milano   Tel. (02) 36.50.83

Editrice e Libreria Herder
Piazza Montecitorio 120
00186 Roma   Tel. 679.46.28
                          Fax: 678.47.51

Libreria Hoepli
Via Hoepli 5
20121 Milano   Tel. (02) 86.54.46
                          Fax: (02) 805.28.86

Libreria Scientifica
Dott. Lucio de Biasio 'Aeiou'
Via Coronelli, 6
20146 Milano   Tel. (02) 48.95.45.52
                             Fax: (02) 48.95.45.48

JAPAN – JAPON
OECD Tokyo Centre
Landic Akasaka Building
2-3-4 Akasaka, Minato-ku
Tokyo 107   Tel. (81.3) 3586.2016
                             Fax: (81.3) 3584.7929

KOREA – CORÉE
Kyobo Book Centre Co. Ltd.
P.O. Box 1658, Kwang Hwa Moon
Seoul   Tel. 730.78.91
                             Fax: 735.00.30

MALAYSIA – MALAISIE
University of Malaya Bookshop
University of Malaya
P.O. Box 1127, Jalan Pantai Baru
59700 Kuala Lumpur
Malaysia   Tel. 756.5000/756.5425
                             Fax: 756.3246

MEXICO – MEXIQUE
OECD Mexico Centre
Edificio INFOTEC
Av. San Fernando no. 37
Col. Toriello Guerra
Tlalpan C.P. 14050
Mexico D.F.   Tel. (525) 528.10.38
                             Fax: (525) 606.13.07
E-mail: ocde@rtn.net.mx

NETHERLANDS – PAYS-BAS
SDU Uitgeverij Plantijnstraat
Externe Fondsen
Postbus 20014
2500 EA's-Gravenhage   Tel. (070) 37.89.880
Voor bestellingen: Fax: (070) 34.75.778

Subscription Agency/   Agence d'abonnements :
SWETS & ZEITLINGER BV
Heereweg 347B
P.O. Box 830
2160 SZ Lisse   Tel. 252.435.111
                             Fax: 252.415.888

**NEW ZEALAND –
NOUVELLE-ZÉLANDE**
GPLegislation Services
P.O. Box 12418
Thorndon, Wellington   Tel. (04) 496.5655
                             Fax: (04) 496.5698

NORWAY – NORVÈGE
NIC INFO A/S
Ostensjoveien 18
P.O. Box 6512 Etterstad
0606 Oslo   Tel. (22) 97.45.00
                             Fax: (22) 97.45.45

PAKISTAN
Mirza Book Agency
65 Shahrah Quaid-E-Azam
Lahore 54000   Tel. (42) 735.36.01
                             Fax: (42) 576.37.14

PHILIPPINE – PHILIPPINES
International Booksource Center Inc.
Rm 179/920 Cityland 10 Condo Tower 2
HV dela Costa Ext cor Valero St.
Makati Metro Manila   Tel. (632) 817 9676
                             Fax: (632) 817 1741

POLAND – POLOGNE
Ars Polona
00-950 Warszawa
Krakowskie Prezdmiescie 7   Tel. (22) 264760
                             Fax: (22) 265334

PORTUGAL
Livraria Portugal
Rua do Carmo 70-74
Apart. 2681
1200 Lisboa   Tel. (01) 347.49.82/5
                             Fax: (01) 347.02.64

SINGAPORE – SINGAPOUR
Ashgate Publishing
Asia Pacific Pte. Ltd
Golden Wheel Building, 04-03
41, Kallang Pudding Road
Singapore 349316   Tel. 741.5166
                             Fax: 742.9356

SPAIN – ESPAGNE
Mundi-Prensa Libros S.A.
Castelló 37, Apartado 1223
Madrid 28001   Tel. (91) 431.33.99
                             Fax: (91) 575.39.98
E-mail: mundiprensa@tsai.es
Internet: http://www.mundiprensa.es

Mundi-Prensa Barcelona
Consell de Cent No. 391
08009 – Barcelona   Tel. (93) 488.34.92
                             Fax: (93) 487.76.59

Libreria de la Generalitat
Palau Moja
Rambla dels Estudis, 118
08002 – Barcelona
                 (Suscripciones) Tel. (93) 318.80.12
                 (Publicaciones) Tel. (93) 302.67.23
                             Fax: (93) 412.18.54

SRI LANKA
Centre for Policy Research
c/o Colombo Agencies Ltd.
No. 300-304, Galle Road
Colombo 3   Tel. (1) 574240, 573551-2
                             Fax: (1) 575394, 510711

SWEDEN – SUÈDE
CE Fritzes AB
S–106 47 Stockholm   Tel. (08) 690.90.90
                             Fax: (08) 20.50.21

For electronic publications only/
Publications électroniques seulement
STATISTICS SWEDEN
Informationsservice
S-115 81 Stockholm   Tel. 8 783 5066
                             Fax: 8 783 4045

Subscription Agency/Agence d'abonnements :
Wennergren-Williams Info AB
P.O. Box 1305
171 25 Solna   Tel. (08) 705.97.50
                             Fax: (08) 27.00.71

Liber distribution
Internatinal organizations
Fagerstagatan 21
S-163 52 Spanga

SWITZERLAND – SUISSE
Maditec S.A. (Books and Periodicals/Livres
et périodiques)
Chemin des Palettes 4
Case postale 266
1020 Renens VD 1   Tel. (021) 635.08.65
                             Fax: (021) 635.07.80

Librairie Payot S.A.
4, place Pépinet
CP 3212
1002 Lausanne   Tel. (021) 320.25.11
                             Fax: (021) 320.25.14

Librairie Unilivres
6, rue de Candolle
1205 Genève   Tel. (022) 320.26.23
                             Fax: (022) 329.73.18

Subscription Agency/Agence d'abonnements :
Dynapresse Marketing S.A.
38, avenue Vibert
1227 Carouge   Tel. (022) 308.08.70
                             Fax: (022) 308.07.99

See also – Voir aussi :
OECD Bonn Centre
August-Bebel-Allee 6
D-53175 Bonn (Germany)   Tel. (0228) 959.120
                             Fax: (0228) 959.12.17

THAILAND – THAÏLANDE
Suksit Siam Co. Ltd.
113, 115 Fuang Nakhon Rd.
Opp. Wat Rajbopith
Bangkok 10200   Tel. (662) 225.9531/2
                             Fax: (662) 222.5188

**TRINIDAD & TOBAGO, CARIBBEAN
TRINITÉ-ET-TOBAGO, CARAÏBES**
Systematics Studies Limited
9 Watts Street
Curepe
Trinidad & Tobago, W.I.   Tel. (1809) 645.3475
                             Fax: (1809) 662.5654
E-mail: tobe@trinidad.net

TUNISIA – TUNISIE
Grande Librairie Spécialisée
Fendri Ali
Avenue Haffouz Imm El-Intilaka
Bloc B 1 Sfax 3000   Tel. (216-4) 296 855
                             Fax: (216-4) 298.270

TURKEY – TURQUIE
Kültür Yayinlari Is-Türk Ltd.
Atatürk Bulvari No. 191/Kat 13
06684 Kavaklidere/Ankara
                             Tel. (312) 428.11.40 Ext. 2458
                             Fax : (312) 417.24.90
Dolmabahce Cad. No. 29
Besiktas/Istanbul   Tel. (212) 260 7188

UNITED KINGDOM – ROYAUME-UNI
The Stationery Office Ltd.
Postal orders only:
P.O. Box 276, London SW8 5DT
Gen. enquiries   Tel. (171) 873 0011
                             Fax: (171) 873 8463

The Stationery Office Ltd.
Postal orders only:
49 High Holborn, London WC1V 6HB
Branches at: Belfast, Birmingham, Bristol,
Edinburgh, Manchester

UNITED STATES – ÉTATS-UNIS
OECD Washington Center
2001 L Street N.W., Suite 650
Washington, D.C. 20036-4922 Tel. (202) 785.6323
                             Fax: (202) 785.0350
Internet: washcont@oecd.org

Subscriptions to OECD periodicals may also be
placed through main subscription agencies.

Les abonnements aux publications périodiques de
l'OCDE peuvent être souscrits auprès des
principales agences d'abonnement.

Orders and inquiries from countries where Distribu-
tors have not yet been appointed should be sent to:
OECD Publications, 2, rue André-Pascal, 75775
Paris Cedex 16, France.

Les commandes provenant de pays où l'OCDE n'a
pas encore désigné de distributeur peuvent être
adressées aux Éditions de l'OCDE, 2, rue André-
Pascal, 75775 Paris Cedex 16, France.

                                          12-1996